TRAUMA AND EMERGENCY
HEALTH CARE
MANUAL

TRAUMA AND EMERGENCY HEALTH CARE MANUAL

A GUIDE FOR NURSING AND MEDICAL STUDENTS

WOMBEOGO AND KUUBIERE

authorHOUSE®

AuthorHouse™ UK Ltd.
1663 Liberty Drive
Bloomington, IN 47403 USA
www.authorhouse.co.uk
Phone: 0800.197.4150

Published by AuthorHouse   04/07/2014

ISBN: 978-1-4918-9780-5 (sc)
ISBN: 978-1-4918-9781-2 (e)

Authors' contact

Dr. Michael Wombeogo (PhD (UK), FWACN (Nigeria), FGCNM (Ghana), IPN
Specialist (Belgium), MA (RSA), MSc (UK), MA (UK), BA (Ghana), ADPM (RSA),
PGC (RSA), SRN (Ghana), Head Clinical Skills

Dr. Callistus B. Kuubiere (FACH (Germany), MBCHB (Ghana), BSc (Ghana), Head
Human Biology Dept and Trauma &Orthopaedic Consultant)
School of Medicine and Health Sciences (SMHS)
Box 1350
Tamale Campus
University for Development studies (UDS), Tamale, N/R. Ghana
Tel: 0242388584/0261115892
Email: mwombeogo@yahoo.com/mwombeogo@gmail.com

DEDICATION

This book is dedicated to all nursing and medical students of School of Medicine and Health Sciences (SMHS), Tamale Campus, University for Development Studies (UDS).

About the Book

This book entitled Trauma and emergency health care manual: a guide for nursing and medical students is a training tool used at School of Medicine and Health Sciences (SMHS) skills laboratory to train nursing and medical students on the basics of emergency health care and management. The expatiation on key emergency or first aid areas such as motor accidents, clothing and helmet removal, functions of a First aider, wounds and fracture management, sepsis and management of infections, resuscitation for both adults and children, poisons, stings and bites are essential for emergency practices in nursing and medical training in Ghana. Emergency care for nurses and medical doctors is a big challenge in the various health facilities in Ghana as almost on daily bases health professionals are confronted with one emergency situation or another due to road traffic or domestic accidents.

This book is meant to provide a fundamental learning tool for nursing and medical students who happen to be front line medical team members during emergencies.

CONTENTS

Foreword...xiii
Acknowledgement.. xv

Chapter 1   General Principles of Trauma and
                Emergency Health Care...............................1
- Introduction to Emergency health care.....................1
- What is an accident? 2
- Steps followed when administering First aid in
  emergency situations 2
- Response during minor ailments...............................3
- The importance of first aid in Emergency health care 3
- Qualities of emergency health care giver.....................4
- Limitations...5
- Precautions taken when offering
  Emergency health care 5
- Measures to prevent or minimise cross infection 5
- Examination of a casualty at the emergency scene......6
- Assessment..6
- The recovery position..7
- History taking and physical examination 8
- History Taking steps 9
- When casualty is conscious 9
- When casualty is unconscious,
  look for external clues..9
- Table 1 Possible signs and symptoms 10

Chapter 2   Techniques and Procedures in
                Emergency Health Care...............................12
- Clothing removal technique......................................12
- Removing clothing in upper body injuries 12
- Removing clothing from lower body injuries 13

- Box 1 Removing an open face helmet 13
- Box 2 Removing a close or full helmet or headgear ...14
- Monitoring vital signs 15

Chapter 3 Principles of Emergency Wound Care 17
- Introduction 17
- What is a wound? 17
- Classification of wounds 18
- Background to the fundamentals of
  the wound healing process 20
- The clinical (normal) healing mechanism 21
- Factors influencing wound healing........................25

Chapter 4 Principles of Wound Treatment 29
- Introduction 29
- The objects of wound treatment 29
- Methods of treating wounds 30

Chapter 5 Essentials of Anti-Sepsis and Asepsis 33
- Introduction 33
- Antisepsis..34
- Asepsis 35
- Rapid hands disinfection in emergencies 36

Chapter 6 Bleeding from Special Regions of the Body 37
- Introduction 37
- Bleeding from the mouth..37
- Bleeding from the nose (Epistaxis) 39
- Bleeding from the lungs..41
- Bleeding from the urinary tract...............................42
- Bleeding from the rectum 42
- The oesophagus and stomach regions 43
- Bleeding from the ear (the acoustic meatus) 44
- Bleeding under the skin 45

Chapter 7 Principles and Management of Shock 46
- Introduction 46
- Causes 46
- Characteristics..47
- Treatment...47
- Types of shock and their treatments........................48
- Diagnosis and treatment 51

Chapter 8   Principles and Management of
Asphyxia and Resuscitation.....................................54
- Causes 54
- Signs and symptoms..54
- Treatment methods and procedures 55
- Heimlich Manoeuvre 55
- Resuscitation..56
- Phases of the terminal state....................................57
- Disturbances of the organ function during the
terminal state..57
- Manual resuscitation techniques 58
- Emergencies requiring CPR 63
- Artificial respiration 65
- Introduction 65
- When artificial respiration is used..........................65
- Principles behind exhaled air resuscitation..............65
- Human intervention techniques..............................66
- Mouth-to-mouth (kiss of life)................................67
- Disadvantages of mouth-to-mouth resuscitation 67
- Mouth-to-nose technique.......................................68
- Advantages of exhaled air resuscitation (EAR)........68
- Sylvester Brosche method 68
- The Heimlich manoeuvre 69
- Procedure for Heimlich manoeuvre 69

Chapter 9   Principles of Managing and
Caring for Burns and Scalds 71
- Introduction 71
- Degrees of severity of burns or scalds.....................72
- Depth of Burns 73
- Superficial burns..73
- Partial-thickness burns...74
- Full-thickness Burns 74
- Extent of Burns..75
- In a healthy adult:...75
- Severe Burns and Scalds...75
- Complications..76
- Assessing a Burn 76
- Treatment...77

- Severe Burns 78
- Minor Burns 78
- Burns to the Mouth and Throat.................................79
- Treatment of Burns to the Mouth and Throat.........79
- Minor Burns and Scalds...79
- Treatment of Minor Burns and Scalds 80

Chapter 10 Principles of Managing Fractures.......................81
- Causes 81
- Symptoms 81
- Classification of fractures 82
- Treatment of fractures in emergency situation 83

Chapter 11 General Principles and
            Causes of Unconsciousness....................................88
- Fainting (Syncope)..91
- Infantile convulsions 93

Chapter 12 Poisons and Management in Poisoning.................95
- Introduction 95
- Types of poisons...96
- Food poisoning...98
- Prevention methods 101

Chapter 13 Stings and Bites..102
- Signs and symptoms..102
- Treating Bee, Wasp and Hornet Stings...................102
- Emergency management..103
- Scorpion sting 104
- Signs and symptoms..104
- Treatment..104
- Bites 104
- Treating Spider Bites 105
- Black Widow Spider Bite 105
- Brown Recluse Spider Bite 105
- Non-Venomous Stings and Bites 106

Bibliography 109

FOREWORD

This book entitled Trauma and emergency health care manual: a guide for nursing and medical students, is a training tool used at School of Medicine and Health Sciences (SMHS) skills laboratory to train nursing and medical students on the basics of emergency health care and management. The expatiation on key emergency or first aid areas such as motor accidents, wounds and fracture management, poisons, stings and bites are essential for emergency practices in nursing and medical training in Ghana. The production of this book has been provoked by the burning concern of nursing students at SMHS of UDS to facilitate learning and skills acquisition in this direction.

We are sincerely grateful to the students and staff of the SMHS particularly those who encouraged me to produce this book.

Dr. Michael Wombeogo

ACKNOWLEDGEMENT

We acknowledge the support of our families, friends and students of School of Medicine and Health Sciences (SMHS) of University for Development Studies (UDS). The authors of books, journals and articles from which we made references to, we duly acknowledge you for your impressive pioneering work in this subject matter.

This book would not be made possible without the support and cooperation of the Vice Chancellor and the Dean of SMHS and the Head of Allied Health Sciences department. We acknowledge the support of Alhaji I.K Antwi, University Librarian of UDS for proof reading and helping with the publication of this book.

Above all we are highly humbled for the grace and love the Almighty God bestowed on us to facilitate the production of this book. Amen!!!

The Authors

CHAPTER 1

GENERAL PRINCIPLES OF TRAUMA AND EMERGENCY HEALTH CARE

Introduction to Emergency health care

The growing publicity in recent times associated with fatal accidents on the roads, in the factories, at the domestic environment, at school and at playgrounds, is arousing the public conscience and increasing the demands for emergency health care training needs.

Emergency primary health care often referred to as "First Aid", is a combination of emergency medical measures used in the event of a sudden illness or accident which is aimed at saving or sustaining life till appropriate medical/surgical attention is available to enhance it.

Emergency primary health care is a broad phenomenon of health strategies applied in an immediate and coincidental circumstance in order to avert further suffering, regression of disease condition, danger, and eventual loss of life. Unlike general emergency health care, *First Aid* is all other forms of assistance available at the instance of the first aid giver and offered for any form of illness or accident pending the arrival of or access to definitive medical assistance.

Persons who offer emergency health care to victims vary and include the following categories:

1. *Unskilled person* - services rendered by a layman or a non professional in First Aid or emergency care, who often has

none of the necessary means, equipment or medical care support materials to perform such a function.

2. *Qualified person* - services rendered by a specially trained health worker (health assistants, physician assistants, nurses, laboratory technician and First Aiders).

3. *Medical doctor* - one who has instruments, apparatus, medicines, blood and blood substitutes, etc. at his/her disposal to perform or carry out emergency care services.

What is an accident?

An accident is an injury to body parts causing impairment of their functions as a result of an unexpected effect of the environment. It is often difficult to access medical service promptly after an accident and so the care rendered on the spot or before the victim is moved to medical facilities is of extreme importance.

Accident victims and relatives in extreme desperation often seek help at the nearest medical institution such as chemist's shop, the dental clinic, laboratory, epidemiological station or even a kindergarten whose workers must render help immediately.

Steps followed when administering First aid in emergency situations

1. Remove the external hazard (electrical current, high or low temperature, compression heavy objects) or remove the victim from the hazardous environment (water, burning premises, or premises with accumulated poisonous gases) before commencing first aid treatment.

2. Render aid to victims based on the seriousness and type of injury, accident, or illness (e.g. arresting bleeding, dressing an open wound, artificial respiration, cardiac massage, or administration of antidotes).

3. The victim or casualty must be immediately transported to the nearest health facility, when stable or while stabilising and call National Ambulance Services

Response during minor ailments

Emergency first aid should not be confined to the treatment of only more serious injuries such as severe bleeding and broken bones, but should include minor ailments. Examples of minor ailments demanding prompt and decisive attention include

- Bruises
- Abrasions
- Strains
- Sprains
- Dog-bites
- Cat-bites
- Scorpion stings
- Bee and other insects stings

When these emergencies are neglected or handled with less seriousness, may have far reaching and unpleasant consequences.

The importance of first aid in Emergency health care

- *To safe life* by ensuring

  1. Airway is patent
  2. Breathing is ongoing and regular
  3. Circulation is stabilised
  4. Haemorrhage is controlled and casualty is not in direct contact with offending object
  5. The general adverse effects of the injury are controlled.

- *To prevent further injury or complication* by skilfully treating

  1. Injuries, illness or mental conditions
  2. Dazed and frightened victims needing counselling to prevent exposure to further risk.

- *To effect improvement in the health of the victim,* or at least ensure that the condition does not worsen, by providing

1. Comfort
2. Reassurance
3. Shelter and
4. Effective care for the victim

- *To prepare for medical treatment,* good first aid is designed to form a basis for subsequent specialised treatment by the clinician.
- *To assist the clinician* by supplying details of the accident, injury and first aid treatment given progress report and giving further help as may be needed under professional supervision

Qualities of emergency health care giver

- *Organising ability,* which includes,

  1. Control of bystanders
  2. passer-bys
  3. obtaining swift and effective help where necessary

- *Self-confidence,* ability to remain well composed in a distressful and dramatic situation
- *Empathy/Sympathy,* when dealing with an injured and frightened person whose world has suddenly been turned upside down a few words of sympathy, reassurance and encouragement are of the greatest possible value as an aid to restoration of morale and recovery.
- *Tact and patience* are needed in dealing with harassed friends and relations, well-meaning and less-informed onlookers and occasionally unimaginative officials.
- *Judgement and thought,* Emergency health care cannot only be learned from text books nor can it be practised by the rule of the thumb. An intelligent appreciation of the situation is essential, for no two accidents are alike nor do injured persons respond in the same way.

Limitations

It must be clearly understood that Emergency health care as a first aid activity has its limitations. Any attempt at more ambitious treatment may easily prove harmful to the patient and throw the whole activity into disrepute. Anyone who undertakes the treatment of a minor accident assumes responsibility for any actions taken in that regard. For instance, do not attempt to remove a piece of grit from an injured eye if you are not a professional in eye care and treatment at the eye theatre.

Precautions taken when offering Emergency health care

1. *Protect yourself, the casualty and bystanders* from injury or death by first identifying risk factors around the accident scene
2. Assess the safety of the *situational environment* and the resources available to you and the kind of help you may need before treating a casualty.
3. Ensure that the *accident environment is safe*, for instance, turning off a live switch, putting off flames, or stopping all on-coming vehicles that may have potential or actual danger to life.
4. *Remain calm* and take some few deep breaths to assist you recall your first aid knowledge and procedures before proceeding with treatment, but swiftly.
5. Then carry out a quick survey of the casualty to *determine the level of consciousness* and what action to take immediately to safe life.
6. Ensure that cross infection from you to a casualty or from one casualty to the other or from a casualty to you is highly minimised, particularly when you are treating open wounds.

Measures to prevent or minimise cross infection

○ Simple hand washing with water and soap before and after attending to each casualty
○ Wear disposable gloves (if available)

○ Cover all skin abrasions on your hands or any uncovered part of the body closest to the body of the casualty with gauze and elastic plaster

○ Avoid as much as possible direct contact of any open wound, among others

7. Dispose soiled materials cautiously and professionally to prevent cross infection and injury to another person using approved disposal measures and procedures.

8. Manage after-stress with tact especially following treatment of casualties in road traffic accident situations.

Examination of a casualty at the emergency scene

Assessment

The first line of action when called to an emergency scene is to assess the following

- Potential risks to yourself and the casualty (remember your safety first before you can deal with any danger to the casualty or bystanders)
- The casualty's health state and level of consciousness, whether there are signs of life or he responds to voices or speech, gentle touch, shaking or deep tap
- Seek assistance from bystanders or call for help when people are not within reach
- Place casualty in recovery position and loosen tied clothing (see next page for a detail exposition on the recovery position),
- Open the airway to allow easy respiration
- If respiration is shorting, pulse is dull and level of alertness is undeterminable, give the casualty 30-40 chest compressions, followed by 2-3 rescue breaths (if casualty is an adult) or 4-5 rescue breaths (if the casualty is a child), and continue for at least 1-2 minutes.
- Continue with chest compressions and rescue breaths till medical assistance is available.

The recovery position

This position helps a semiconscious or unconscious person breathe and permits fluids to drain from the nose and throat so they are not choked.

If a casualty is unconscious but is breathing and has no other life-threatening conditions, they should be placed in the recovery position.

Significance of the recovery position

Putting someone in the recovery position will ensure

- The airway remains open and clear.
- That any vomit or fluid will not cause them to choke.
- That respiration will take place systematically with less difficulty

Caution

DO NOT PLACE A CASUALTY IN THE RECOVERY POSITION IF

- you think the person may have a spinal injury,
- back or neck injury

Therefore do the following

- do not move them,
- Place your hands on either side of the casualty's face and gently lift his jaw with your fingertips to open the airway.
- Take care not to move the neck.
- However if breathing is or becomes noisy tilt the head slightly backwards without over extending the neck in order to allow more air to enter the nose and mouth into the lungs (that is no tongue falls back to obstruct airways).
- Maintain the back and neck in position until help arrives

Checklist of recovery position procedure

- Make sure casualty is not wearing glasses.
- Put the legs of the casualty straight.
- Place the nearest arm to you at right angles to the body of the casualty when kneeling next to him/her.
- Then make sure the other arm is across the casualty's chest;
- Make sure the back of his/her hand is held against his/her nearest cheek.
- Hold the casualty's thigh that is furthest from you with your other hand and pull the knee up.
- Keep casualty's foot flat on the ground.
- Pull down slowly the casualty's knee that is raised and gently roll him/her towards you.
- Keep casualty's hip and knee at right angles by moving the upper leg slightly. (in order not to allow casualty roll back onto his/her face.
- Tilt the head gently backwards to keep the airway open.
- Maintain a clear and clean environment for fresh air to freely reach the casualty

History taking and physical examination

After ascertaining that casualty is out of imminent danger and initial judgement has been made as to what to do with and for the casualty, it is necessary to find out what happened by taking historical antecedent of the incident from the casualty and primary bystanders (those who have been at the spot prior to the occurrence of the incident and witnessed the start and current state of it). Then perform a physical examination to determine the extent of injuries incurred by the casualty, in order to commence any appropriate treatment or seek for immediate assistance.

History Taking steps

When casualty is conscious

- Ask the casualty of his name, where he lives and where he was going at the time
- Find out when he last had a meal
- Establish past illnesses whether he is on any medications (side effects of some medications can be responsible for sudden falls causing significant injuries)
- Find out if casualty took any alcoholic beverage prior to the accident
- Establish the severity of the fall and the extent of injury to the site of the body casualty fell on
- Establish the environment where the incident occurred (in hot water, hot and humid or cold room, exposed to wind, rain, electric circuit, etc.)
- If possible, document findings for onward transmission to the clinician later on.

When casualty is unconscious, look for external clues

- (alcohol, drugs, auto-injectors—possible anaphylactic shock, diabetic patients; inhalers—indications for asthmatics, SOS Talisman or Medic Alert bracelet)
- Presence of live electric cables near by
- Robes tied on a tree and hanging downwards (sign of suicide attempts)
- Observe the extent and type of injuries sustained as a result of the fall or crash
- Observe some indicators to the type and severity of the force of impact
- Pay attention to bystanders reports on incident and how the casualty sustained the injuries.

Table 1 Possible signs and symptoms

Mode of identification	sign	Symptom
The casualty's own experiences		Pain, heat, cold, thirst, nausea, tingling sensation, weakness, dizziness, abnormal sensation, etc
Subjective observation (seeing and describing)	Anxiety and painful expressions, burns, sweating, abnormal skin colour, vomit, bleeding from orifices, response to touch, swelling, muscle spasm, loss of normal movement, circumstantial evidences (containers, medicines, hammers, etc.)	
Feeling	Dampness, abnormal body temperature, abnormal pulse rate, swelling, grating bone ends, etc.	
Hearing	Noisy and laboured respiration, groaning, sucking sounds, response to touch and speech, crepitus	
Smell	Alcohol, chars from burning source, gas or fumes, solvents or glue, urine, faeces, cannabis (wee), perfumes, etc.	

Physical examination of a casualty from head-to-toe (Cleaver et al, 2006:34)

- Examine scalp for swelling, bleeding or depression (take care not to move casualty when neck injury is suspected)
- Communicate by word into both ears of casualty to ascertain whether he can hear and observe for any discharges from the ears
- Examine both eyes and note eye opening, pupil reaction to light, their size (level of equality), any foreign object, blood or bruising in the cornea or sclera of the eye
- Check nose for any discharges
- Take note of the state of respiration (rate, depth and nature of breath sounds)
- Observe colour of lips and inspect mouth for any dentures which could be dislodged and cause respiratory difficulties
- Observe for any cyanosis (colour, temperature and state of skin)
- Loosen clothing around the neck and look for signs of SOS Talisman or medical warning medals or bracelets
- Ask casualty to take a deep breath (if conscious); while feeling gently with both hands the abdomen, observe the chest expansion distribution
- Feel gently the clavicles and shoulders for any deformity, irregularity or tenderness
- Check for the level of flexibility of elbows, wrists and fingers by gently bending and straightening the arm and observe for any abnormal sensations and colour in the limbs
- If there is any loss of sensation or impairment, think of a possible spinal injury; if so, do not move casualty; seek for professional assistance (if possible); immobilise casualty before lifting or moving from site to a safer place for management.
- Examine abdomen for any possible bleeding or tenderness
- Examine hips, legs, knees and ankles and observe any changes in shape, colour, fracture or bleeding
- Document findings clearly
- Transport casualty to the nearest health facility for further management
- Report findings recorded at the accident scene to health authorities

CHAPTER 2

TECHNIQUES AND PROCEDURES IN EMERGENCY HEALTH CARE

||||||||||||||||||||||||||||||||||||||||||||||||||||||||||||||||||||||||||||||||||||||||||||||||||||||||||||||||||||||||

Clothing removal technique

Unnecessary removal of clothing should be avoided because too much exposure increases chilling and shock. When removing clothing is essential, the first aider must act tactfully and prevent exposure of the body as far as possible. Seek permission from client if possible before cutting clothing. When removing clothing, make sure you do not cause undue discomfort to the injured part as much as possible (*See Cleaver et al, 2006: 40-41*).

Removing clothing in upper body injuries

- Loosen fastens such as buttons, zips, hooks, and gently slip garment off shoulders of casualty
- Remove garment from the uninjured arm side first and then to the injured side from the sleeve
- Maintain the injured arm in position while removing the garment off it, making sure not to cause undue pain or further injury
- In the case of clothing that cannot be unfastened (sweaters, poly shirts and round neck T-shirts), commence from the uninjured side by easing the sleeve off it. Then roll up the garment while stretching the neck of the garment over the head of the casualty. Finally, slip off the garment over the injured arm, taking care not to cause further injury or pain to the arm

- In a situation where clothing is stuck to injured part of the body, a similar process should be followed, starting from the uninjured side to the injured side. However, when garment reaches the injured side, the garment could be split opened at the seams if practicable to remove it without further cause of pain or injury to casualty.

Removing clothing from lower body injuries

- Communicate cordially with casualty and enquire from him whether he feels pain at the leg or foot.
- Unfasten laces or strips as may be applicable.
- Support leg at the ankle joint with an open palm of one hand and carefully remove the shoe. In the case of long boots, it is advisable to slit them open at the back of the seam with a knife or scissors.
- Remove socks by pulling them off gently from the uninjured site and off at the injured end. Where this is not possible, gently slip through the socks with your two fingers and raise the sock away from the body of the casualty. Then use scissors to slit through the sock to open it.
- Gently pull up the bas of trousers to expose the calf and knee so as not to let the trouser end come into contact with the injured foot. If the thigh must be exposed, pull trouser down from the waist, first unfastening belt, zip or buttons.

Box 1 Removing an open face helmet

In case of a motor traffic accident where casualty has a protective crash helmet on, do not remove it unless

1. *it's presence interferes with treatment of the casualty and might endanger life*
2. *cause of accident and injury are well established not to have had any effect on the head or neck regions*
3. *upon a clear judgement, it is affirmed that removal is absolutely necessary*

The following steps can be followed to remove an open face helmet or head gear

- Establish level of consciousness (awareness or alertness) by communicating to the casualty
- Seek the cooperation of the casualty to carry out the removal procedure
- Unfasten or cut the chinstraps. Then support the casualty's head and neck with both hands making sure that the head and neck are in alignment with the spine
- Seek the assistance of a bystander or a fellow aid worker to grip the lateral ends of the helmet with both hands, while standing at the end of the casualty and facing you at the opposite side of the casualty.
- The helper gently but firmly pulls off the helmet from the head to take off pressure from the head
- Then he/she (the helper) should gently lift the helmet upwards and backwards to completely free it from the head.

Box 2 Removing a close or full helmet or headgear

In case of a motor traffic accident where casualty has a full face protective crash helmet on, do not remove unless

1. *Its presence interferes with treatment of the casualty and might endanger life*
2. *Cause of accident and injury are well established not to have had any effect on the head or neck regions*
3. *upon a clear judgement, it is affirmed that removal is absolutely necessary*

Procedure is as follows:

- Establish level of consciousness (awareness or alertness) by communicating to the casualty
- Seek the cooperation of the casualty before you carry out the removal procedure

- Unfasten or cut the chinstraps. Then support the casualty's head and neck with one hand and hold the mandible firmly. Then ease your fingers under the base of the rim of the helmet. Then ask a bystander of a helper to hold the helmet with both hands from the back of the head of the casualty.
- Ask the helper to tilt the helmet backwards and gently upwards to free the base of the helmet off the casualty's chin.
- Maintain support of neck, head and mandible, while the helper tilts the helmet forwards slightly so as to pass it over the base of the skull. Then helmet is lifted straight off the casualty's head

Monitoring vital signs

Vital signs such as level of response (eye movement, motor activity, response to stimuli), pulse, blood pressure, respiration and temperature are very essential in helping you identify specific changes in a casualty.

Response

Use ASPU technique

- Alertness
- Speech or voice response
- Pain response
- Unresponsive to stimuli

Pulse

- Brachial pulse (inner part of upper arm, mostly for children)
- Radial pulse (below the wrist creases at the base of thumb)
- Carotid pulse (side of neck at the hollow between thorax and the sternocleidomastoid muscle)
- Temporal pulse (medial part of the thigh, just below the inguinal ligament
- Check the pulse rate, strength and rhythm

Respiration

- Normal—adult: 12-16 breaths/min; children: 20-30 breaths/min
- Check for the following—rate (# of breaths per minute); Depth (deep or shallow breath sounds); Ease (easy, difficult or painful breaths); Noise (quiet, noisy and types of noise)

Temperature

- Use thermometer (digital—tongue or armpit (30 seconds), mercury—tongue, armpit, anus (2-3 minutes), forehead—use for a child (30 seconds), or ear sensor—use for a very sick child (1 sec), etc)
- Feel exposed body to ascertain level of body heat
- Normal body temperature - 36.1-37.1°C
- High temperature, ≥ 37.5°C (Pyrexia) which could be due to infection
- Low temperature ≤ 35.9°C (hypothermia) may be due to exposure to cold or wet conditions

CHAPTER 3

PRINCIPLES OF EMERGENCY WOUND CARE

Introduction

Intact skin and mucous membrane support optimal body function, yet disruption may occur in either of the two whenever an applied force is greater than the tissues can withstand. Both accidental and intentional breaking of the skin and mucous membrane happens frequently. There arises the need to reflect on the ubiquitous nature of wounds (that is when there is a disruption in tissue integrity with or without opening of skin) in order to realise that the body has remarkable ability to recover. It is the recovery ability of the body of the body as well as measures supporting the healing process that are the focus of this lecture.

Intact skin and mucous membrane protect the internal body environment against invasion by infectious organisms and loss of body fluid in excess. Thus any wound predisposes the individual to the potential for infection and fluid loss. In order to allow the human organism to heal itself as usual, all actions related to healing or repair process should give the body optimal support so that it can combat those organisms that have already entered and conserve body fluids.

What is a wound?

A wound or open injury is an abnormal break of the continuity of the skin, mucous membranes, deeply lying tissues, or surface of an internal organ as the result of a mechanical or some other action or force. A cavity is likely to form between tissues as a result of the penetration of the body with a penetrating (wounding) object. This type of cavity is often called *a wound canal.*

Classification of wounds

In traumatology, distinctions are made between complex and noncomplex wounds and between wounds in relations to the continuity of skin covering (either open or closed). Wounds can also be categorised according to their *cause,* (intentional or accidental), *type of wound,* such as *vulnus scissum* (incised wound), *vulnus contusum* (contusion), or *vulnus conquaessatum* (crushing wound), and *presence or absence of pathogenic micro-organisms.*

A complex wound involves not only the skin, the subcutaneous, and muscles, but also large lymphatic and blood vessels, nerves, or bones. A non-complex wound generally does not extend beyond the subcutaneous. Open wounds communicates with the environment and disrupt the connection between all skin layers and include incised wounds, penetration wounds, puncture wounds, bites, perforating wounds, and lacerations.

A wound may either be closed or open. In a *closed wound,* no break in skin continuity can be observed. A closed wound may be caused by a direct blow with a blunt instrument or by usual straining, twisting, or sudden deceleration might precipitate a closed wound. An *open wound* is one in which there is disruption of the skin and mucous membrane. This type may be caused by a sharp blow or object. The open wound allows direct loss of fluid from the body and the entrance of foreign particles and organisms into the body. Exudates collect with resultant swelling, loss of function and pain accompany both closed and open wounds. Inherent in both injuries is the potential for wound complications as discussed later in the proceeding paragraphs.

Wounds may occur *intentionally* (surgical) or *accidentally* (traumatic). When a surgeon operates, there is a disruption of skin and mucous membrane. This *intentional type* of wound is usually performed under special conditions, with sharp instruments, and the extent of the wound is related specifically to the intended purpose and its edges, readily approximated. An accidental wound is unexpected, frequently has ragged (jagged) edges and occurs under septic conditions. These factors alter the repair mechanism and the rate of healing and accordingly provide a greater potential for complications.

Descriptive terms may be used to refer to the types of wounds. An *abrasion* (excoriation) is a superficial wound caused by scraping or sliding of skin surface directly over a firm, fixed surface. The floor burn caused by sliding on a sidewalk or floor with skin surface exposed is an abrasion.

A *contusion* (*bruised wound*) usually does not involve a break in the continuity of skin surface but underlying damages include breaking of blood vessels and swelling as a result of *haematoma* formation caused by blood thrombosis to the affected localised area. The discolouration caused by the extravasations of blood into tissue is called *ecchymosis*.

An *incised wound* is usually caused by a sharp cutting instrument such as a knife, razor, and scalpel blade particularly during surgical operations. The edges of an incised wound are sharp, clean, smooth and straight. Symptoms of an incised wound are dehiscence (separation of the edges of the wound), bleeding (diffuse, venous, or arterial) and pain

Lacerated wounds are characterised by ragged, irregular and torn and crushed edges and are caused by sharp and heavy instruments (e.g. axe, sword). These types of wounds have the danger of affecting bones.

Penetrating wounds are those caused by piercing instruments through skin and mucous membrane and into deeper tissues and organs. If the instrument or object enters and exits from the deeper tissue or organs, it is referred to as a *perforating wound*. For instance, gunshot wounds are classified as perforating wounds. Bullet goes straight through the body forming two orifices, the inlet and the outlet, blind when the bullet remains in the body, or tangential, when the bullet grazes the body and causes only superficial injury.

A *punctured wound (stab)* is produced by a sharp pointed object piercing deep tissue, leaving a very small opening on the surface. It is mostly caused by such instruments such as a knife, bayonet or a needle. The small superficial openings can be deep, but with a narrow wound canal and as a rule is disrupted and becomes zigzag-like because the tissues pull apart due to muscular contraction or skin movements. This makes punctured wounds especially dangerous since it is difficult to determine the depth of the injury or whether the internal organs are wounded. Latent damage to the internal organs may give rise to internal haemorrhages, peritonitis (inflammation of the peritoneum), or pneumothorax (a condition when air penetrates the pleural cavity). Besides, there is a potential of entry of anaerobic organisms (*clostridium tetani or Clostridium welchii*) during the time of injury.

Finally, wounds may be classified as clean, contaminated, or infected, depending on the presence or absence of pathogenic organisms. Any break in the continuity of the skin or mucous membrane presents a risk for micro-organisms entering the wound and multiplying. In addition, any collection of exudates poses a depot for the growth of micro-organisms. A *clean wound* contains no pathogenic micro-organisms, because of the aseptic conditions under which surgical incisions are made; these wounds are usually considered clean. A *contaminated wound* is one concurring in a manner in which there is a great likelihood of pathogenic micro-organisms invading the wound. **Every wound, except for cuts made by sterile instruments under sterile conditions during an operation, should be considered contaminated**. An *infected wound* is one in which pathogens have invaded and overcome the body's first line of defence, producing clinical signs of infection (cardinal signs of inflammation). The infected wound may be referred to as a *septic* wound.

Background to the fundamentals of the wound healing process

The process of wound repair is a normal reaction to injury and the keystone on which modern surgery is founded. The issues of scar formation, sepsis and eventually the healing mechanism spelled out confrontations and misunderstandings in the reparative mechanism until recent times.

In the pioneering days of de Chauliac, Pare and Lister wound infections occurred with such regularity as to be considered a routine phase of wound healing. The appearance of pus was regarded as a sign that eventual repair might be expected and when Semmerlweiss, Lister and others worked to show that sepsis had an adverse effect on healing they were ostracised for their unorthodox thinking. Thankfully, surgical innovations have delved into new techniques particularly with antiseptic skin preparation and the advent of antibiotics, now make sepsis the exception rather than the rule.

To a lesser extent surgeons are still grappling with the issue of scarring. Just as the early surgeons expected sepsis, surgeons today accept scarring. Some scarring is a biological necessity, though certain operative sequelae such as keloid, hyper-tropic scars, puncture marks and cross-hatching from skin sutures should be avoidable. Progress in this field requires an understanding of the normal process of wound repair and the best setting for this is a healthy patient with a normal vascular system.

The clinical (normal) healing mechanism

It is convenient to describe the local process of wound healing in four stages, although in reality, this is a continuous process with one stage merging into the next.

Stage 1: The phase of wounding and traumatic inflammation (0-3 days)

> ➤ This is the time of injury—the moment at which tissue disruption occurs. Within minutes there will be loss of organ function, haemorrhage and subsequent blood clotting, bacterial contamination, foreign body contamination and demonstrates all the features of acute inflammation due to infection.

➤ Depending on the degree of contamination and virulence of the invading bacteria, the amount of haemorrhage, the healing time increases or minimises. Redness, swelling and local heat (calor) occur in a recent wound as part of the normal healing process. This phase begins within a few minutes of wounding and lasts for about 3 days.

➤ When tissue is disrupted, blood vessels are injured and bleed into the space created. Platelets and the coagulation system cause blood to clot the wound. Injured blood vessels thrombose and bleeding stops.

➤ Damage tissue and mast cells secrete histamine and other enzymes causing vasodilatation of surrounding capillaries and exudation of serum and white cells into the damaged area.

➤ This increased blood supply with oedema and engorgement of surrounding vessels accounts for the inflammatory appearance, warmth and throbbing sensation experienced by the patient.

➤ This is a beneficial auto-reaction and there is no need to attempt to cool the area or reduce the swelling unless it occurs in a closed compartment where important structures may be compressed (as in the neck or closed to facial compartments).

➤ At this stage, two important cell types arrive in the wound. Polymorphoneucleocytes and macrophages which combine in defence against bacteria and clear all debris, damaged tissue and blood clot in the wounded site and hence begin the process of repair.

Box 3 Inflammation and the reparative process

If *too little inflammation* occurs, the reparative process is *slow* (e.g. after steroid administration or in debilitating disease state). If *too much*, the repair process is *prolonged* due to excessive exudation of cells competing for the limited nutrition at the wounded site. Prolonged inflammatory phase leads to i) delayed formation of new tissue, ii) retarded development of tensile strength and iii) bacteria multiply at the site

➤ A heightened inflammatory reaction can sometimes be avoided by removing predisposing factors such as

Box 4 Predisposing factors in the inflammatory reaction

i) traumatised or devitalised tissue
ii) foreign bodies
iii) excess suture material
iv) poor handling technique

Stage II: The destructive phase (2-5 days)

➤ Polymorphs and macrophages[1] clear the wound of devitalised and unwanted material

➤ Healing stops when macrophages are eliminated, but the process of healing continuous unabated even with a major reduction in the number of polymorphs in a clean wound.

➤ Macrophages are very important in wound repair because they act as "*director cells*" attracting further macrophage migration and multiplication of fibroblasts—the cell which synthesises collagen. Collagen is the body's principal structural protein and can be found in fresh wounds as early as the second day.

➤ Macrophages therefore play a significant role of recruiting fibroblasts and by clearing unwanted debris.

➤ New blood vessels then grow into the wound from the edges and fibroblasts follow them to assume their role, which is clearing the wound of unwanted material.

➤ Increase cellular activity and enzymatic breakdown of unwanted fibrin and dead cells is brought about as a result of macrophage and polymorph lysosomal content discharged into the area.

➤ Increase osmolality to the area attracting more water by osmosis through the action of the small protein molecules produced by enzymatic degradation.

➤ This will culminate in an obligatory swelling of the part.

➤ This may pose **some problems**, for instance, following a fracture of the tibia where the muscles, nerves and blood

---

1 Macrophages are large mobile cells with the ability to engulf and digest bacteria and dead tissue. They are found in the best and worst vascularised tissues and in tissues spaces such as peritoneal cavity.

vessels are contained within a restrictive fascial compartment (or POP) and ischaemic numb leg that may require fasciotomy.

Angiogenesis:

Endothelial cells create a capillary network. This entire process leads to the formation of granulation tissue

Stage III: The proliferative (fibroplasias) phase (3-24 days)

➢ The real healing begins here and the initial scab forms at this time

➢ Fibroblasts, endothelial and epithelial cells multiply

➢ Fibroblasts line up behind the macrophages

➢ Then produce the strands of collagen, which is the main constituent of skin, tendons, ligaments, bones, cartilage, fascia and scar tissue. It occurs at about the 5th-7th day. The production of collagen by fibroblasts happens best in an acid environment. The most prominent stimulants are vitamin C and lactate ions which accumulate in areas of low oxygen such as in the centre of the wound. So, without vitamin C, collagen synthesis is inhibited while breakdown continues.

➢ Vitamin C deficiency (scurvy) will cause **i) purpura** (breakdown of unsupported blood vessels with bleeding) and ii) **failure of new wounds to heal.**

➢ These chemical and cellular activities results in granulation tissue

➢ Three terms frequently used to describe particular healing phenomena are *eschar, proud flesh and keloid*. Eschar is the sloughing tissue resulting from gangrene, corrosive trauma and thermal burns. During the healing of large surface wounds, *proud flesh* may develop. This refers to the formation of excessive amounts of oedematous, soft granulation tissue. A *keloid* is a progressively enlarging scar which is raised, tumour-like and triangular. This overgrowth of scar tissue results from excessive collagen formation in the skin's dermal layer during the repair of connective tissue.

Stage IV: The maturation phase (24 days-1 year)

> Decrease in vascularity of the scar, shrinkage of the fibroblasts, enlargement and reorientation of the collagen fibres and augmentation of tensile strength (about **50%** within the third week) occur here;
> The epithelial covering of the wound becomes multilayered, cell production becomes balanced by cell death and collagen hydrolysis, degradation and absorption occur.
> This is when the dusky red appearance of vascular granulation tissue changes to the pale white avascular scar tissue
> Increased tension also causes more collagen to be laid down and if the healing wound is examined under the microscope at this stage, the overall effect appears to be one of lacing the wound edges together with a three-dimensional weave.

In addition to the four cellular and chemical phases described above, there are two other mechanisms to be considered and these include contraction and epithelisation.

Contraction is the process by which large wounds become small without the need for secondary closure or skin graft (amputation above the knee, for instance) as a result of what is commonly described as the *motive force* through the action of myofibroblasts.

Epithelisation is the process by which squamous epithelial cells shed and are replaced from below through the combined activities of multiplication, flattening and migration towards the area of cell defect. Successful epithelisation occurs only if the cumulative effects of physical manipulation, drying, bacterial enzymes, wound area and limitation of blood supply do not exceed the capacity for available cells to divide and move across the surface.

Factors influencing wound healing

> *The extent of the injury* - a small and superficial injury heals more readily than an extensive, deeper wound.

➤ *The type of tissue injured* - certain tissues heal by regeneration (epithelial tissue on squamous surfaces of skin, interior of the bucal cavity, vagina and cervix, lining of salivary glands; vascular epithelium, parts of cornea and kidney, epithelium of the digestive and respiratory tracts. Others tissues such as nerves, myocardial tissue, brain tissue and renal glomeruli do not regenerate.

➤ *The nutritional state of the individual* - healing rate may be retarded and incidence of complications increased in persons with inadequate amounts of vitamin C, in particular, minerals and calories. Deficiencies of these occur in the malnourished or underfed individual and in surgical patients who have been starved just before and after operation. Vitamin C is essential in collagen formation and its absence will lead to a weak wound. Zinc is essential for wound healing and its absence can delay wound repair. Protein is essential for new tissue formation and its absence predisposes the individual to infection. Thus in poorly nourished and debilitated patients, parenteral hyper-alimentation may be prescribed as an aid to wound healing.

➤ *Oxygen* - all wounds are inherently hypoxic due to blood vessel interruption at the site of the injury. Without oxygen, collagen formation is reduced and slow. Thus hypovolemia (reduced blood volume through excessive blood loss), local oedema causing constriction of blood vessels, firm bandaging or splinting causing pressure on regional blood vessels, atherosclerotic vascular changes, myocardial insufficiency.

➤ *The presence of pathogenic micro-organisms* - all open wounds provide direct pathways for pathogens to enter the body; under aseptic conditions the normal inhabitants of the naso-pharynx, skin and gut threaten to enter the wound. Under contaminated conditions the organisms waiting to enter are many and include *staphylococcus aureus* which produces coagulase causing thrombosis of blood vessels and further haemorrhage at the site and subsequent necrosis of tissue.

➤ *Concomitant diseases* - such as marked hypertension, diabetes mellitus, uraemia, renal acidosis, liver disease, carcinomas, etc.

> *Chemotherapeutic agents* - anti-inflammatory agents such as cortisone which is associated with nitrogen and potassium depletion will eventually cause a reduction in collagen formation; aspirin inhibits platelet aggregation and subsequent capillary oozing; immunosuppressive and cancer drugs such as steroids, antimicrobials, cytotoxic drugs, etc, have a depressive effect on bone marrow and reduce available blood cells.
> *Stress response* - retention of sodium and fluid, excretion of potassium and nitrogen occur during the stress response and may have varying effects on wound healing. Both the severity and the extent of the response alter the result.
> *Radiotherapy agents/methods*

Approaches:

Wounds heal in one of four ways. The manner in which healing occurs varies with the individual involved, the location of and type of wounding.

1. Primary wound healing (sanitatio per primam intentionem or *first intention)*: if the wound heals without infection or separation of wound edges, the process is called healing by first intention or by primary union. This will result in a very small scar and requires atraumatic, precise approximation of the wound edges and the absence of disturbances in wound healing, such as infection. This is the desired approach for clean surgical wounds in which the wound edges lie neatly close together.

2. Secondary wound healing (sanitatio per secundam intentionem or *second intention)*: when wound edges are not approximated, granulation tissue fills in the opening prior to healing and this form of healing is referred to as healing by second intention. With this approach, the wound is left open and nature is allowed to run its course through the inflammatory phase, proliferation phase, and differentiation phase. This approach to wound healing is highly safe and should be used if there is uncertainty regarding the viability of the wound area based on the previously mentioned local and general factors. An example is bite wounds caused by humans or animals.

3. *Third intention*: when there is a combination of the above, namely wound edge initially left open and later approximated or initially sutured and later broken open, healing may be described as occurring by third intention
4. Delayed primary wound closure occurs for wounds in which uncertainty exists regarding the viability of tissue or the degree of contamination or infection. Examples include wounds older than six hours or large, heavily contaminated wounds. As soon healthy, well vascularised granulation tissue is visible in the wound, the edges of the wound can be brought together. Hermetic sealing of the wound is avoided to minimise the risk of infection.

Table 2 Stages of the healing process

Characteristic	First intention	Second intention	Third intention
Wound edges	approximated	Not approximated	Initially not approximated, later approximated
Infection	Absent	Frequently present	Frequently present
Granulation tissue	Small amount	Large amount	Large amount
Scar	Small	Large	Large
Healing rate	Short	Long	Long
Example	Surgical incision	Decubitus ulcer	Eviscerated surgical wound

This process of healing is orderly and systematic; although individual differences and complications developing during the process lead to variations in the time the process takes.

CHAPTER 4

PRINCIPLES OF WOUND TREATMENT

Introduction

The type of wound, its site and the effect of the wound on the body influence the kind of treatment to be given. The amount of treatment supplied in any particular case depends on the available facilities and resources and the environment of the accident. In street accidents, for instance, attempts should be made to stop severe bleeding immediately and treatment for shock commenced simultaneously. However, the wound itself should be treated quickly and cautiously, with more professional treatment to be continued at the health centre or at the hospital.

The objects of wound treatment

- Haemorrhage arrest
- Prevention and treatment of shock
- Sepsis and contamination prevention
- Healing and recovery promotion\

These objects are of equal importance geared towards improving the welfare and complete recovery of the casualty. Consequently, they should be treated in the order given as a matter of preference.

In emergency wound treatment, the extent of treatment is highly dependent on

- The facilities available to the Emergency health care giver at the time
- The environment of the accident (for instance, in street accidents, severe bleeding must be controlled immediately while treatment for shock follows (if any). Treatment of the wound should be done quickly, just to protect the wound from external infection and reduce pain. More thorough treatment should be postponed and done at the hospital or at the nearest health facility.
- The number of casualties involved needing treatment

Methods of treating wounds

- Routine
- Temporary

Routine method

This is the best method of treatment, preferably in suitable conditions when maximum amount of facilities and resources are available to do so. This method should be performed in the following circumstances

- All minor wounds (cuts, scratches, bruises, etc.)
- Small wounds that will require further medical attention later
- Larger wounds under medical supervision
- Larger wounds where medical attention is less likely in about 6 or more hours (treat shock and other related complications, while routine treatment is delayed until medical attention is available.

Routine treatment

- Control haemorrhage
- Place casualty in suitable position (blood escapes with less force when the casualty is in the lying position, for this reason, the casualty should be made to lie flat in the case of severe bleeding. Casualty may sit up in a case of less bleeding and shock).

- Patient care, which includes reassurance and sympathetic encouragement
- Expose wound by removing clothing or boots; this is significant as to give the care giver a fair idea of the severity of the bleeding. In the event that bleeding is severe, and the care giver is faced with limited or no dressing materials, part of the casualty's clothing should be left on the wound to act as a partial protection against the entry of germs pending actual treatment at the hospital later.
- Cleanliness; in the process of offering Emergency health care, the care giver is likely to contaminate his hands with blood and sand and should therefore wash his hands thoroughly with soap and water before proceeding with any further treatment. This is significant to prevent cross infection and cross contamination or infection of the wound.
- Expose wound completely following a thorough examination of wound, type of haemorrhage that is occurring, possible presence of foreign bodies. The risk of cut tendons must be borne in mind. If no fractures, raise injured part and support it in an elevated position, in order to reduce haemorrhage.
- Remove loose foreign bodies in the depth of the wound and not those deeply embedded in order not to precipitate further haemorrhage. The emergency care giver must not probe the wound for foreign bodies; this should only be done at the health centre by the doctor and specialised health professionals.
- Cleanse the wound with a suitable antiseptic solution (hydrogen peroxide 10%, Milton, Dettol 1 in 20, and Cetavlon) soaked in small pieces of cotton wool, dabbed over the surface, squeezed so that the antiseptic penetrates to the depths of the wound. It is also used to cleanse the skin around the wound. Take care not to contaminate the wound with dirt derived from the skin.
- Further examine the wound during wound cleansing and seek medical attention if there is need to deal with bleeding or need to insert stitches.

- Apply antiseptic solution if wound is trivial and does not need urgent medical attention. Suitable antiseptics include surgical spirit, undiluted dettol, or weak iodine solution.
- Apply a suitable dressing using gauze or lint placed with the smooth sides next to the skin.
- Provide support to the wounded limb in order to relieve pain and prevent secondary shock.
- Arrange for medical assistance since the injections of preventive drugs may be desirable to avoid the risk of tetanus and other complications.

NB: *When medical facilities are nearer, or there is a limited opportunity of seeking medical attention in good time, it is advisable to cover the wound temporarily with a field or standard dressing. Arrest haemorrhage; cleanse the skin round the wound with surgical spirit or any suitable antiseptic available.*

CHAPTER 5

ESSENTIALS OF ANTI-SEPSIS AND ASEPSIS

||||||||||||||||||||||||||||||||||||||||||||||||||||||||||||||||||||||||||||||||||||||||||||||||||||||||||||

Introduction

In time past, Louis Pasteur, a French scientist proved that putrefaction and fermentation were as a result of the actions of micro-organisms. Joseph Lister, an English surgeon just after Pasteur's discovery concluded that wound infection was caused by micro-organisms. However, before Lister's conclusion, the Russian scientist Nikolai Pirogov intimated that wound infection resulted from what he described as "hospital miasma" and employed alcohol, silver nitrate and iodine to disinfect wounds.

Human beings daily activities expose them to huge number of micro-organisms in the air, surrounding objects, skin and mucous membranes of healthy persons. When there happen to be a break in the continuity of the intact skin or mucous membrane due to wounds, burns, abrasions, punctures, weakened body defence forces by circulatory disorders, cooling, exhaustion or any general ailment, micro-organisms can easily gain access to the body and cause purulent-inflammatory foci (suppuration, abscesses, phlegmons). In more severe cases, micro-organisms enter the body and can cause general toxaemia or sepsis.

During surgical interventions (operations, IM/IV injections, blockages, etc.), the continuity of the intact skin is disrupted, thus enabling infection to penetrate the body. The process of infection control and prevention in wounds is known as *antisepsis*, while maintaining an infection free wound is known as *asepsis*.

Antisepsis

According to Buyanov (1985), antisepsis is a combination of measures aimed at killing micro-organisms in the wound and preventing them from getting deep into the tissues. This is accomplished by mechanical, physical, chemical and biological means.

Mechanical antisepsis - removal of dead and crushed tissue, blood clots and foreign bodies from a wound (e.g. debridement of a wound).

Physical antisepsis - involves exposure of the wound to infrared radiation, introduction into the wound of drains, tampons and sterile hot pads or towels to induce the outflow of pus and fluid in order to hinder the development of wound infection. This method is usually employed when first aid is rendered by medical personnel

Chemical antisepsis - otherwise termed disinfectants include the following

- **Hydrogen peroxide solution** (sol. hydrogenii peroxidati); colourless liquid, weak disinfectant, used in a 3% solution, mostly for skin and mucous membranes and foams when in contact with a dirty wound.
- **Potassium permanganate** (Kalii permanganas); weak disinfectant, used in a 0.1-0.5% solution to disinfect purulent wounds, as a tanning agent in burns, ulcers and bedsores
- **Boric acid** (Acidum Boricum); used as a 2% solution to wash mucous membranes, wounds and body cavities
- **Tincture of iodine** (Tinctura Jodi); disinfect operative field and hands of surgical personnel, treat contaminated wounds, lacerations and scratches.
- **Iodonate** (iodonatum); used in a 1% solution for disinfecting operative field and hands in emergency situations.

- **Iodoform** (Tri-iodomethane); comes in powder form and employed in ointments and emulsions to disinfect purulent wounds.
- **Chlorame B** (Chloraminum B); an antiseptic and deodorant, comes in 1-2% solution, used to irrigate putrefactive wounds, while 0.25-0.5% solution is used to disinfect hands, gloves and instruments.
- **Mercuric chloride or perchloride** (Hydrargyri dichloridum); used in 1:1000 dilution to disinfect contaminated articles and soiled gloves. It is a strong poison and should be stored away from easy contact with intact or broken skin.
- **Silver nitrate** (Argentum nitricum); has a cauterising and anti-inflammatory action, weak solutions of 1:3000 and used to douche the urinary bladder, wound granulations in 10-30% solution.
- **Ethyl alcohol, ethanol** (Spiritus aethylicus); as 70-96% solution, it is used to disinfect cutting instruments (scalpels, scissors), suturing materials (silk) and the operative field.

Biological antisepsis

These include antibiotics (substances produced by micro-organisms themselves or synthetically prepared to heighten the organism's own defence forces (vaccines, sera, gamma globulins, etc.). antibiotics was first studied and obtained by the Russian scientist Z.V. Ermolyeva. The commonly used antibiotics include penicillin, monomycin, erythromycin, chloramphenicol, tetracycline, gentamycin sulphate, kanamycin, etc.

Antibiotics are used orally, IM/IV, subcutaneously, in solutions to wash and irrigate wounds and in ointments and emulsions for dressing.

Asepsis

This is a process by which pathogenic micro-organisms are prevented from entering objects and wounds. It is maintained by disinfecting everything that comes into contact with the wound. Sterilisation is the process by which bacteria and their spores on operation linen, suturing material, surgeons' gowns, gloves and hands are rendered completely of all forms of germs. And object is considered sterile when its surface and interior are free of microbes capable of reproducing. Sterilisation can be done in one of the following ways:

- Steam under pressure (autoclaving) at 110°C-120°C
- Dry heat at 140-180°C in 10-15 minutes
- Calcinations (same as dry heat)
- Boiling, is the simplest method of sterilisation for metal instruments, glassware, syringes, gloves, rubber catheters and tubes, plastic instruments; usually at 100°C for 30 minutes twice with a 6-hour interval in between.
- Burning, in the case of emergencies, instruments can be burned in spirit, i.e. a metal basin is used with the instruments immersed in spirit and ignited. The flames provide a reasonable disinfection but do not reliably sterilise the instruments as can be done by autoclaving, etc.
- Keeping contaminated articles in antiseptics and antibiotics
- Radiation (gamma rays), ultraviolet radiation (mercury quartz lamps)
- Gases

Rapid hands disinfection in emergencies

Before rendering first aid, the hands must be disinfected as far as possible by washing with soap under running water and drying with a clean towel. Then if available, ethyl alcohol, 5% iodine alcohol solution or 5% phenol solution (carbolic acid) and a solution of mercuric chloride (1:1000), etc. If sterile gloves are available they can be won over non-sterile hands.

CHAPTER 6

BLEEDING FROM SPECIAL REGIONS OF THE BODY

Introduction

Haemorrhage does occur in any part of the body following and injury to the intact skin or mucosal lining. However, when bleeding occurs in some vital areas of the body, it indicates an instance of urgent concern and the emergency care giver is therefore called to action without any further delay. These special regions of First Aid importance include

- The mouth
- The nose
- The lungs
- The urethra
- The rectum
- The oesophagus and stomach regions
- The ear
- The skin

Bleeding from the mouth

Causes

- Direct—1) wounds such as lacerations of the gums, tongue, cheeks, inner lips, etc. 2) from the socket of a tooth

- Indirect—1) from the nose, 2) from the lungs, 3) from the stomach

NB: the seat of haemorrhage can be determined through careful examination of the mouth and underlying structures

Lacerations of the gums, tongue, cheeks, inner lips, etc

These occur as a result of a direct injury to the mouth, such as a knock on the mouth or as a result of an epileptic fit. Bleeding is usually of capillary origin and can be easily controlled using simple measures such as

- Let casualty be in the sitting position with head forwards and tilted slightly towards the injured side
- give ice to suck, or
- cold water to hold in the mouth
- rinse with a mild antiseptic like weak dettol solution
- discourage swallowing, for swallowing could induce nausea and vomiting
- In the case of severe lacerations bleeding can be controlled by applying direct pressure on the affected part. The wound should be covered with a piece of lint before applying any pressure.

Bleeding from the socket of the tooth

This kind of bleeding occurs after a tooth extraction. The bleeding originates from the capillaries. Pack the tooth socket with an antiseptic pack wearing disposable gloves, where professional assistance is available. Caution patient not to spit or wash out his mouth. This is important so as not to disturb the blood clots forming which aids the natural arrest of haemorrhage.

In the event of delay in obtaining professional assistance, the emergency care giver should

- Cover the socket with a small roll of lint or gauze, over which a cork or any hard object placed and patient asked to bite on the pad.
- Then apply a bandage from the chin across the head and tied tightly on top of the scalp. This is to help the jaw muscles from extreme fatigue if they are to remain clenched for a while.
- Then let patient to be in the sitting position with the head resting backwards.

Bleeding from the nose (Epistaxis)

Haemorrhage from the nose is called *epistaxis*. Bleeding is venous in origin and sometimes nose-bleeding can be very serious with fatal consequences.

Causes

- Rupture of tiny, distended vessels in the mucous membrane, commonly from the anterior septum where three important blood vessels enter the nasal cavity. These are, the anterior ethmoidal artery on the kesselbach's plexus, the sphenopalatine artery in the postero-superior region and the internal maxillary branches located at the back of the lateral wall
- Trauma - blows on the nose may cause bleeding, broken nose
- Infection - naso-pharyngitis, sinusitis,
- Drugs
- Cardiovascular diseases - blood dyscrasias (e.g. haemophilia), leukaemia
- Hypertension - common among the elderly who are most likely to suffer apoplexy
- Nasal tumours - polyps
- Varicose veins - these when ruptured causes profuse haemorrhage
- Physiological growth - common in childhood and adolescence and during the process of development
- Low humidity

- Altitude - ascending high altitudes like mountain climbing
- Foreign bodies - commonly in children
- Vigorous nose blowing and nose-picking

Management

- Reassure the casualty that nose bleeding is not dangerous and that the haemorrhage can be controlled sooner in order to allay his anxiety on the sight of blood.
- Seat casualty on a chair or position in the sitting position with his head slightly forward
- Loosen tight clothing round the neck, chest and waist to permit return of venous blood towards the heart
- Instruct casualty to

  1. keep his mouth open and breathe through the mouth
  2. avoid blowing the nose which might dislodge any blood clot formed in the process
  3. avoid swallowing, coughing, sniffing or spitting

- keep casualty warm to make him comfortable, without over heating
- apply pressure on the nostrils in the following manner

  1. pinch the nostrils firmly with the thumb and forefinger against the septum of the nose for 10 minutes or until the bleeding has stopped
  2. if bleeding still persists, pack the nostril with a wet cotton wad (wetted with hydrogen peroxide, lidocaine, epinephrine or Vaseline) leaving part of it outside the nose
  3. then maintain compression on the nostrils for 10 minutes or more
  4. leave the cotton wad there for a few hours after bleeding stops; then take it out very carefully
  5. once bleeding has stopped clean around the nostrils with a clean wet towel or handkerchief with the casualty still leaning forwards

6. In older persons, bleeding may come from the back part of the nose and cannot be stopped by pinching it. In this case have the casualty hold a cork or corn cob or something similar and suitable between his teeth and leaning forward, sit quietly and try not to swallow until the bleeding stops. The cork helps him from swallowing and that gives the blood a chance to clot.

- Seek medical attention as soon as possible

Bleeding from the lungs

Bleeding from the lungs or *haemoptysis* occurs in diseases such as tuberculosis. In haemoptysis;

- the blood has a characteristic bright red colour and frothy because it is mixed with air
- blood is coughed up
- previous history of lung infection is probable

Causes

- diseases—acute pneumonia, TB
- fractures to ribs and lungs
- drowning

Management

- Reassure and keep patient calm and collected; coughing up of blood often causes fright and anxiety in casualties and should be encouraged to realise that the haemorrhage will soon respond to treatment.
- Keep casualty relaxed preferably in a well ventilated but darkened room to enhance complete body and mental relaxation
- Keep casualty in lying position with shoulders slightly raised and turned towards the affected side, supported by pillows and with the head hanging down (this position aids coughing

and prevents flooding of the unaffected parts of the lungs with blood).
- Make casualty comfortable as much as possible
- Do not give any stimulants as they tend to increase bleeding
- Give ice to suck, which sometimes aids the arrest haemorrhage
- Then seek medical attention

Bleeding from the urinary tract

Bleeding from any part of the urinary tract is called *Haematuria*

Causes

- Disease
- Injury to any of the genito-urinary system
- Bladder stones

Management

- Reassure the casualty
- Keep casualty relaxed for a while
- Advise casualty to go to the nearest health centre for further investigation to establish cause of bleeding as soon as possible

Bleeding from the rectum

Bleeding from the rectum, particularly from the higher level of the digestive tract which produces a characteristic stool termed *malaena* often due to bacterial digestion of blood. Frank blood may be seen if the source is below or near the anus.

Causes

- Local injury to rectum itself
- Complicated fracture of the pelvis
- Disease—peptic or duodenal ulcer, haemorrhoids (piles), helminthias, carcinomas of the gastrointestinal tract (GIT)

Characteristics

- Rectum—blood is red colour in the stool
- Higher digestive system—dark brownish stool
- Signs and symptoms of shock

Management

- Reassurance
- Complete rest
- Position casualty in the lying position with the foot end of the bed raised by 15-20 centimetres using bed blocks or improvised with books or cement blocks, etc.
- Seek medical aid

The oesophagus and stomach regions

Bleeding in the oesophagus or in the stomach becomes visible when vomiting or regurgitation of food occurs. The vomiting of blood is termed *haematemesis*

Causes

- Diseases of the stomach, e.g. gastric ulcer,
- Varicose veins at the lower end of the oesophagus
- Previously swallowed blood from nose-bleeding, fracture of the base of the skull or jaw
- Sudden stress can trigger gastric ulcer bleeding
- Carcinoma of the oesophagus

Signs and symptoms

- Signs and symptoms of shock (giddiness, pallor, cold, faintness, quick and feeble pulse)
- History of dyspepsia (indigestion)
- Visible blood

- Peculiar characteristic (clotted dark red blood mixed with digested food in the vomitus or coarse dark brown in appearance when food is partially digested)

Management

- Reassurance—often times the casualty does not realise that he is vomiting blood and when reassuring him about his condition, all reference to blood should be avoided.
- Complete bed rest in a comfortable position, preferably in the lying position with the head turned to one side and supported during the act of vomiting. Avoid anything per os, except ice to suck as can be tolerated.
- Keep casualty lying down as far as possible with the feet raised and head and shoulders low by raising the end of the bed with bed blocks about 15-20cm placed at the casters. Remove all pillows.
- Obtain medical care at the earliest possible time

Bleeding from the ear (the acoustic meatus)

Causes

- Trauma or lacerations due to blows, heavy fall on the ear
- Scratches
- Fracture of the cranial bones (at the back of the skull)

Management

- Reassure casualty
- Position casualty in a comfortable position
- Apply a light dressing over the ear and bandage it in position
- Avoid removing objects seen or suspected to have caused the bleeding
- Seek urgent medical attention for further investigation and treatment

Bleeding under the skin

Bleeding under the skin is termed *haematoma*

Causes

- Trauma, e.g. a blow, heavy fall, etc.
- Contusion

Management

- Reassure casualty and keep him calm
- Apply any evaporating lotion on the swelling (or contused part), or
- Apply a firm pad over the swelling
- Secure pad in position by a tight bandage

CHAPTER 7

PRINCIPLES AND MANAGEMENT OF SHOCK

Introduction

Shock, is a state of acute circulatory insufficiency of the blood. Put differently, shock is a failure of the circulation to adequately supply the tissues with adequate oxygen and nutrients. It is the result of the inability of the heart to pump adequate volume of blood at sufficient pressure for normal blood passage through the major organs and tissues of the whole body.

Causes

- heart failure—as a result of coronary arterial occlusion with acute myocardial ischaemia, trauma to heart structures, pericarditis, cardiac tamponade, toxaemia (bacterial or viral) and drug intoxication (overdose of pethidine).
- injury
- Severe burns—can lead to hypovolaemia from excessive plasma loss
- reduction in blood volume (oligaemia or hypovolaemia) from haemorrhage
- Major surgery, for instance, multiple trauma case
- Sudden infection or poisoning
- abnormal extracellular fluid volume resulting from excessive loss of water, diarrhoea, cholera

- Electrolytes (ionic conductors) from the gastrointestinal tract, kidneys, or skin.

Characteristics

- Apathy
- Weakness,
- shallow breathing,
- rapid heartbeat,
- feeble pulse,
- lowered blood pressure, and
- Coldness and clamminess of skin.

During the early stage, consciousness is retained, but alertness is diminished. Sudden peripheral circulatory failure, however, affects the brain and causes fainting. In less severe shock compensatory constriction of the blood vessels helps restore circulation, but if shock persists, compensatory mechanisms fail and local anaemia damages vital organs, such as the brain, heart, liver, kidneys and lungs in a vicious cycle of events.

Treatment

First aid for shock includes

- Having the victim lie down, keeping him or her warm but not overheated,
- Stopping any bleeding, and, if the person is not breathing, maintain patent airways and administer artificial respiration, with cardiac massage support.
- Complete cardiac arrest can be an inevitable sequel unless the heart is restored within 3-5 minutes. External cardiac massage (ECM) is commonly used to stimulate the heart to action. A physician or paramedic may administer oxygen and adrenergic drugs and take measures to lower the body temperature if it is very high. Medical personnel may give intravenous salt solutions (normal saline, ringers lactate,

colloids) and transfusions of plasma or whole blood may be required when shock is due to blood loss.

- In medical centres, treatment is monitored by direct measurements of arterial blood pressure and arterial blood gases. Sometimes blood pressure is increased by vasopressors (drugs that constrict blood vessels).

Types of shock and their treatments

Electric Shock is a serious, sometimes fatal physical injury caused by an accidental flow of electricity through the body. Electric shock administered in a carefully regulated dose can also be an effective medical treatment.

Most injuries from electric shock occur through accidental contact with an exposed wire or other part of a live electrical circuit such as electrical wiring or parts of an electrical appliance. Electric shock occurs less frequently through accidental contact with a high-voltage power line or a lightning strike.

Most cases of accidental electric shock involve a hand or arm in contact with a power source. Because the heart and lungs are close to this point of contact, these are the organs injured most frequently. Although electric shock can paralyze the diaphragm or interrupt nerve impulses that regulate respiration (breathing), death usually results from electricity's effect on the heart. The electrical charge breaks the normal rhythm of the heart and induces *fibrillation* (a rapid, irregular fluttering) of the heart muscle, which prevents the organ from beating normally and pumping blood.

Heat generated by contact with a high-voltage current or lightening causes skin burns where the electricity enters and leaves the body. About 1,000 persons in the United States suffer serious burns from electric shock annually. Electric shock also can cause muscles to contract suddenly, which may propel a person to the ground or across a room. A strong electric shock that is not fatal usually interferes with the function of internal organs near the point of contact.

In giving first aid to an electric-shock victim, a caregiver must not touch the victim with bare hands until the source of electricity has been removed safely or the power source shut off. If the victim is not breathing, **mouth-to-mouth or mask-to-mouth resuscitation** is necessary. If the person also has no pulse, **cardiopulmonary resuscitation (CPR)** should be administered until professional emergency help arrives. Burns should be rinsed with or immersed in cold water, blotted dry, and kept clean and covered until the victim can be examined by a physician. First aid also includes the prevention of shock, a reduction of blood flow to body tissues that can cause increased anxiety; pale, cool, clammy skin; rapid, weak pulse; possible fainting; or in more serious cases, coma or death.

The controlled delivery of an electric shock, called defibrillation, is used to restart the heart after a heart attack, which may result from accidental electric shock that stops the heart from beating properly. It also is used to restart the heart during open-heart surgery, and for the treatment of some mental illnesses, especially acute forms of depression, through a procedure called electroconvulsive therapy (ECT).

Anaphylaxis, or anaphylactic shock, is a sudden, severe allergic reaction that can be life-threatening. Anaphylaxis is most often caused by foods, insect bites or stings, and certain medications. Symptoms usually appear within seconds to 30 minutes after coming into contact with the allergen. They involve the whole body and include difficulty breathing, wheezing, low blood pressure, rash; tightness in the chest, swollen throat, vomiting, and feeling light-headed or passing out. Death, though rare, can result from airway obstruction, respiratory failure, or cardiac arrest. Anaphylaxis should be treated as an emergency. Immediate medical attention is required. The most effective form of treatment is an *epinephrine injection*

Allergic reactions can also be triggered throughout the entire body, rather than in one specific location. Called allergic or anaphylactic shock, this response occurs when many cells throughout the body react simultaneously to an allergen, such as bee sting venom. The person may experience hives or welts on the skin, itching all over the body, asthmatic spasms in the lungs, or a sudden drop in blood pressure (*see* Shock). An additional danger is the possibility of swelling in the throat, tongue, and larynx (voice box), which can close the upper airways and cause fatal choking.

Food allergies are fairly common, but they are poorly understood. When foods are digested and the nutrients are absorbed in the intestine, substances in the food probably stimulate the allergic response. The reaction can occur in the intestine itself, resulting in cramps or diarrhoea; in the skin in the form of eczema; or all over the body, also causing allergic shock. Food allergy, which is an immune response, is often confused with food intolerance, which may cause similar symptoms of gastrointestinal discomfort. Food intolerance, however, is caused by many factors other than an allergen; a person may lack an enzyme to digest a particular food, for instance. The resulting symptoms of food intolerance are not triggered by an immune response.

Researchers have identified a definite genetic predisposition to allergies. For example, if one parent has allergies, there is an increased risk that some of the children will also have allergies, although the children may not be sensitive to the same allergens as the parent. If both parents have allergies, the risk that the children will develop allergies is even greater. The most typical time for allergies to develop is in young adulthood, although allergies can develop in a person of almost any age, even within a few months after birth. Allergies in infants are most commonly associated with foods and viral respiratory infections. For reasons that are not clearly understood, children with allergies tend to outgrow them. The child's body somehow readjusts its response to allergens, even those that cause severe reactions, such as food, drugs (especially penicillin), and stinging insects.

Patients are sometimes surprised by the abrupt onset of allergies in adult life. This can occur when the combination of a person's genetic makeup and a longstanding but unnoticed reactivity to an allergen finally culminates in a detectable disease. This so-called allergic threshold is crossed when allergens finally produce enough reaction in the body to cause detectable symptoms. Several as yet unknown genes may be involved in this process.

Many people think that emotions, such as stress, cause allergies, but most physicians believe this is incorrect. In fact, the opposite may be true: People with allergies live with symptoms that can produce serious emotional upset. For example, a person with asthma may have fears about losing the ability to breathe, or possibly choking to death. Allergies, including asthma, are caused by biological factors, although emotions may aggravate an allergic reaction.

Diagnosis and treatment

Diagnosing and treating allergies is usually performed by an allergist, a physician trained to understand the body's immune response. When visiting an allergist for the first time, a patient is usually asked which substances seem to provoke symptoms and whether other family members have allergies. The doctor conducts a brief physical examination, looking in the nose, eyes, ears, and throat, listening to the chest, and examining the skin.

In many cases the allergist will perform allergy skin tests. These are painless injections or scratches into the surface of the skin with tiny amounts of specific allergens, such as pollens or house dust mites. The injection produces a tiny bump in the skin, no larger than the head of a pin. When a person has a positive skin test response (usually occurring after 15 to 30 minutes), there is a local reaction in the skin. The tiny bump gets slightly bigger, slightly itchy, and is surrounded by some redness caused by dilating blood vessels in the area. This reaction indicates to the allergist that cells in the skin contain specific antibodies to the allergens used in the injection. This reaction provides a quick technique for testing and diagnosing specific allergies. Sometimes blood samples evaluated in a laboratory are used to measure the blood levels of IgE antibodies specific to individual allergens, although skin tests have proven to be more accurate, faster, and less expensive in allergy diagnosis.

For most patients with allergies, medicines are used to begin therapy. Most forms of hay fever are easily managed with antihistamines, which relieve the symptoms, such as itching and sneezing, produced by histamine. Asthma is usually treated with medications taken orally or inhaled in vapour form using a metered-dose inhaler. Asthma medications include bronchodilators (drugs that expand the air passages) and anti-inflammatory steroids, which suppress the immune response that causes airway inflammation. In cases of anaphylactic shock, emergency treatment with an injection of adrenaline, also known as epinephrine, is required. This injection quickly widens blood vessels and opens up constricted airways.

If medicines cannot control allergy symptoms, the allergist may suggest allergen immunotherapy, a series of injections of the offending allergens. These injections, or allergy shots, help desensitize the patient to the allergens, thereby reducing the allergic response. For safety reasons, the injections begin with very small allergen doses, close to the amount used in a skin test. The amount of the allergen injected is increased each week for many weeks, until high doses of injected allergens are reached. The goal of immunotherapy is to produce blocking or neutralizing antibodies that provide a protective response in the cells, preventing allergens from binding to the allergic antibodies. When this blocking is achieved, little or no histamine is released in response to the allergens, and allergic symptoms are reduced or eliminated. Allergen immunotherapy is especially effective in overcoming stinging insect allergies, which, if left untreated, can result in a fatal allergic reaction to an insect bite. Immunotherapy is also effective for severe hay fever and, in some patients, for severe, chronic allergic asthma.

In many cases the best allergy treatment is the removal, if possible, of the offending allergens from the patient's environment. For example, the best way to deal effectively with an allergy to cats is to remove cats from the patient's surroundings, although desensitizing injections containing cat extracts are being tested. Unfortunately, some allergens, such as plant pollens, are impossible to avoid, since they float freely through the air. Contact with pollens can be reduced by keeping windows closed and using air conditioners to filter and cool indoor air. House dust mites, a common allergen, can be minimized by frequent cleaning with safe chemicals.

Food allergies are more difficult to diagnose and treat than other types of allergies. Skin tests are unreliable, and blood tests can be inconclusive. When a particular food is suspect, the patient simply should not eat it. If the culprit food is unknown, the allergist may put the patient on a special diet that eliminates various foods. If symptoms decline, the allergist will reintroduce each of the foods one at a time to help identify which food is the offending al

CHAPTER 8

PRINCIPLES AND MANAGEMENT OF ASPHYXIA AND RESUSCITATION

Asphyxiation occurs when air cannot reach the lungs, cutting off the supply of oxygen to circulating blood.

Causes

- Drowning
- Gas poisoning
- Overdose of narcotics,
- Electrocution
- Choking
- Strangulation
- Cerebral haemorrhages
- Traumatic shock

Signs and symptoms

- Victims may collapse,
- Unable to speak or breathe
- Bluish skin
- Brain death within four to six minutes after breathing ceases unless first aid is administered.

Treatment methods and procedures

Heimlich Manoeuvre

In the case of choking, a procedure known as the *Heimlich manoeuvre* can be used to clear the windpipe or the trachea of food or other foreign objects.

In this *procedure* quick upward thrusts are applied to the victim's abdomen to eject the object blocking the windpipe. The first-aid provider

- *Stands behind the victim* with both arms around the victim's waist.
- One fist is placed slightly above the navel and below the rib cage with the thumb against the victim's body. The other hand is used to hold the fist and apply pressure.
- The abdomen is then pressed quickly inward and upward, forcing air from the lungs to eject the object from the windpipe.
- If the victim is too large to hold while standing, or becomes unconscious, the manoeuvre can be accomplished by laying the person down dorsally or face up and the use the heel of one hand in the same manner as above.
- The person performing the manoeuvre must be careful not to apply pressure on the rib cage to avoid breaking ribs, especially in children and the elderly.
- For obese or pregnant choking victims, the provider's hands should be placed over the lower half of the sternum (breastbone) and pressure applied as described above.

For victims of other types of asphyxiation, the most practical method of artificial respiration is the

- *Mouth-to-mouth technique*—the first-aid provider forcefully exhales air into the victim's lungs after first clearing the airway of any obstruction.

- The provider tilts the victim's head backward by placing one hand under the victim's chin and lifting while the other hand presses down on the victim's forehead.
- At this point, the mouth and airway can be checked for foreign objects, which can be removed with the fingers.
- To begin mouth-to-mouth resuscitation, gently pinch the victim's nostrils together to prevent air from escaping out the nose.
- Take normal breaths, seal the victim's mouth with a pocket mask or mouth, and exhale into the mouth. When performed properly the victim's chest should rise visibly.
- The provider then listens for the victim to exhale; if using a pocket mask, it need not be removed. This process is repeated at a rate of about 12 times per minute (one breath every five seconds) for adults and about 20 times per minute for children, using less pressure and volume for children.
- Once beginning artificial respiration, the first-aid provider should continue until the victim begins to breathe or medical help arrives.

In cases of *drowning*, artificial respiration should be attempted even if the victim appears dead. People submerged in cold water for more than 30 minutes who appeared blue have responded to first-aid efforts and recovered with no brain damage.

RESUSCITATION

Resuscitation is a process of restoring breathing or heartbeat following cessation of respiration or cardiac arrest. In other words, resuscitation is the timely restoration of blood and respiration by using a combination of measures to bring a victim to life from a **terminal state**, such as myocardial infarction, shock, massive blood loss, obstruction of the airways or asphyxia, electrocution, drowning, falling debris, etc. The cerebral cortex is the most sensitive to hypoxia or hypoxaemia (low oxygen content in the tissues and blood). In terminal state, the first functions to be switched off are those of the higher part of the central nervous system, i.e. the cerebral cortex. With cessation of oxygen to the brain, consciousness is lost. If oxygen hunger last beyond 3-4 minutes, cerebral activity cannot be restored and is equally lost. The last brain part to die is the medulla oblongata where the automated respiratory and circulatory centres are located. Irreversible death of the brain finally sets in.

Phases of the terminal state

- The pre-agonal state
- Agony
- Clinical death

In the *pre-agonal state,* consciousness is still retained but it is clouded. Blood pressure (BP) drops to zero, pulse is drastically accelerated and thready, respiration is shallow and difficult and the skin becomes pale.

In the *agony state* the blood pressure and pulse are unaccountable, eye reflexes (corneal, and the pupil's reaction to light) disappear, and respiration is of the swallowing type.

Clinical death is a short-term (3-6 minutes) transitory stage between life and death. Respiration and cardiac activity cease, the pupils are dilated, the skin is cold, and the reflexes are absent. At this stage, the vital functions (cardiac activity, cerebral functions and metabolic activities) can still be restored during the period by resuscitation techniques. Later, irreversible changes in the tissues set in and **clinical death** transforms into **true biological death.**

Disturbances of the organ function during the terminal state

At the terminal state, whatever its cause, gives rise to general changes in the organism. Growing hypoxaemia and the functional disorders of the brain give rise to cardiovascular disturbances. The pumping function of the heart in the pre-agonal state is sharply diminished, and the amount of blood ejected by the heart, that is, the cardiac output, is also decreased. The diminished blood supply to the organs and especially to the brain is responsible for the development of the irreversible changes. However, the heart contractions will continue for some time as it possesses its own automaticity. Because these contractions are inadequate, their effect is too small and as a result the pulse filling decreases and it becomes thready, blood pressure sharply diminishes and soon becomes unaccountable. Later the rhythm of the cardiac contractions become markedly impaired and cardiac activity finally ceases.

Throughout the pre-agonal state, the respiration increases and becomes deeper. During the agonal state, BP drops, and respiration becomes irregular, shallow and finally ceases. This cessation leads to a terminal state.

Some vital organs such as the liver and kidneys respond also to hypoxia. Thus, long term oxygen hunger causes irreversible changes in them. Changes in metabolic activities are displayed by oxidative processes and the accumulation in the body of organic acids (lactic and pyruvic) and carbon dioxide. Consequently, the acid-alkaline equilibrium is impaired. Under normal circumstances the blood and tissues have a neutral pH. The fading oxidative processes throughout the terminal state are responsible for a shift of the reaction into the acid side and acidosis sets in. The more prolonged the period of dying the more pronounced is this shift.

When the organism recovers from clinical death, the first to be restored is cardiac activity, then natural respiration and then the cerebral function. Cerebral cortex functionality follows the disappearance of the acid-alkaline state. However there is likelihood that consciousness may not be regained for a long time even after short-term hypoxia and clinical death.

Manual resuscitation techniques

Cardiopulmonary resuscitation (CPR)—this is a type of respiratory first aid that deals with the heart-lung support system and requires special training. In this procedure, which is used for a person who has had a heart attack or respiratory failure, the reviver alternately breathes for the victim and performs external cardiac massage on the person's chest to keep blood moving through the body.

The technique is more easily done by two trained people, one performing each manoeuvre. If the procedure lasts for a long time, it is best that the two people switch positions to be able to sustain the procedure.

Cardio-pulmonary resuscitation (CPR) procedure

Before starting CPR, check:

- Whether the person is conscious or unconscious
- If the person appears unconscious, tap or shake his or her shoulder and ask loudly, "Are you OK?"
- If the person doesn't respond and two people are available, one should call 911 or the local emergency number and the other one should begin CPR.
- If you are alone and have immediate access to a telephone, call 911 before beginning CPR—unless you think the person has become unresponsive because of suffocation (such as from drowning). In this special case, begin CPR for one minute and then call 911 or the local emergency number.
- If an automatic external defibrillator (AED) is immediately available, deliver one shock if instructed by the device, then begin CPR.

C-A-B

In 2010, the American Heart Association changed its long-held acronym of ABC to CAB—**circulation, airway, breathing**—to help people remember the order to perform the steps of CPR (Mayo Foundation, 2011). This change emphasizes the importance of chest compressions to help keep blood flowing through the heart and to the brain.

Circulation: Restore blood circulation with chest compressions

1. Put the person on his or her back on a firm surface.
2. Kneel next to the person's neck and shoulders.

3. Place the heel of one hand over the center of the person's chest, between the nipples. Place your other hand on top of the first hand. Keep your elbows straight and position your shoulders directly above your hands.
4. Use your upper body weight (not just your arms) as you push straight down on (compress) the chest at least 2 inches (approximately 5 centimeters). Push hard at a rate of about 100 compressions a minute.
5. If you haven't been trained in CPR, continue chest compressions until there are signs of movement or until emergency medical personnel take over. If you have been trained in CPR, go on and check the airway and rescue breathing.

Airway: Clear the airway

1. If you're trained in CPR and you've performed 30 chest compressions, open the person's airway using the head-tilt, chin-lift maneuver. Put your palm on the person's forehead and gently tilt the head back. Then with the other hand, gently lift the chin forward to open the airway.
2. Check for normal breathing, taking no more than five or 10 seconds. Look for chest motion, listen for normal breath sounds, and feel for the person's breath on your cheek and ear. Gasping is not considered to be normal breathing. If the person isn't breathing normally and you are trained in CPR, begin mouth-to-mouth breathing. If you believe the person is unconscious from a heart attack and you haven't been trained in emergency procedures, skip mouth-to-mouth rescue breathing and continue chest compressions.

Breathing: Breathe for the person

Rescue breathing can be mouth-to-mouth breathing or mouth-to-nose breathing if the mouth is seriously injured or can't be opened.

1. With the airway open (using the head-tilt, chin-lift maneuver), pinch the nostrils shut for mouth-to-mouth breathing and cover the person's mouth with yours, making a seal.
2. Prepare to give two rescue breaths. Give the first rescue breath, lasting one second—and watch to see if the chest rises. If it does rise, give the second breath. If the chest doesn't rise, repeat the head-tilt, chin-lift maneuver and then give the second breath. Thirty chest compressions followed by two rescue breaths is considered one cycle.
3. Resume chest compressions to restore circulation.
4. If the person has not begun moving after five cycles (about two minutes) and an automatic external defibrillator (AED) is available, apply it and follow the prompts. Administer one shock, and then resume CPR, starting with chest compressions for two more minutes before administering a second shock. If you are not trained to use an AED, a 911 operator may be able to guide you in its use. Use paediatric pads, if available, for children from ages 1 to 8. Do not use an AED for babies younger than age 1. If an AED isn't available, go to step 5 below.
5. Continue CPR until there are signs of movement or emergency medical personnel take over.

To perform CPR on a child

The procedure for giving CPR to a child age 1 through 8 is essentially the same as that for an adult. The differences are as follows:

- If you're alone, perform five cycles of compressions and breaths on the child, this should take about two minutes, before calling 911 or your local emergency number or using an AED.
- Use only one hand to perform heart compressions.
- Breathe more gently.
- Use the same compression-breath rate as is used for adults: 30 compressions followed by two breaths. This is one cycle. Following the two breaths, immediately begin the next cycle of compressions and breaths.

- After five cycles (about two minutes) of CPR, if there is no response and an AED is available, apply it and follow the prompts. Use paediatric pads if available. If paediatric pads aren't available, use adult pads.

Continue until the child moves or help arrives.

To perform CPR on a baby

Most cardiac arrests in babies occur from lack of oxygen, such as from drowning or choking. If you know the baby has an airway obstruction, perform first aid for choking. If you don't know why the baby isn't breathing, perform CPR.

To begin, examine the situation. Stroke the baby and watch for a response, such as movement, but don't shake the baby.

If there's no response, follow the CAB procedures below and time the call for help as follows:

- If you are the only rescuer and CPR is needed, do CPR for two minutes—about five cycles—before calling 911 or your local emergency number.
- If another person is available, have that person call for help immediately while you attend to the baby.

Circulation: Restore blood circulation

- Place the baby on his or her back on a firm, flat surface, such as a table. The floor or ground also will do.
- Imagine a horizontal line drawn between the baby's nipples. Place two fingers of one hand just below this line, in the centre of the chest.
- Gently compress the chest about 1.5 inches (about 4 cm).
- Count aloud as you pump in a fairly rapid rhythm. You should pump at a rate of 100 compressions a minute.

Airway: Clear the airway

1. After 30 compressions, gently tip the head back by lifting the chin with one hand and pushing down on the forehead with the other hand.
2. In no more than 10 seconds, put your ear near the baby's mouth and check for breathing: Look for chest motion, listen for breath sounds, and feel for breath on your cheek and ear.

Breathing: Breathe for the infant

1. Cover the baby's mouth and nose with your mouth.
2. Prepare to give two rescue breaths. Use the strength of your cheeks to deliver gentle puffs of air (instead of deep breaths from your lungs) to slowly breathe into the baby's mouth one time, taking one second for the breath. Watch to see if the baby's chest rises. If it does, give a second rescue breath. If the chest does not rise, repeat the head-tilt, chin-lift maneuver and then give the second breath.
3. If the baby's chest still doesn't rise, examine the mouth to make sure no foreign material is inside. If the object is seen, sweep it out with your finger. If the airway seems blocked, perform first aid for a choking baby.
4. Give two breaths after every 30 chest compressions.
5. Perform CPR for about two minutes before calling for help unless someone else can make the call while you attend to the baby.
6. Continue CPR until you see signs of life or until medical personnel arrive.

Emergencies requiring CPR

Sudden infant death syndrome

Breast-feeding appears to decrease the risk of SIDS, apparently because it helps prevent respiratory, gastric, and intestinal illnesses, infections, and certain immune disorders that may make infants more susceptible to SIDS. Eliminating smoking around the baby also decreases risk. A doctor may recommend the use of a heart and respiratory monitor for babies at high risk for SIDS. This machine sounds an alarm when the baby stops breathing or when the heart rate is too high or low. Infant caretakers are encouraged to learn cardiopulmonary resuscitation (CPR) in case an infant stops breathing. SIDS is a devastating event for parents, who need support and reassurance for many months afterwards. SIDS support groups, comprised of other parents who have had similar experiences, can be particularly valuable.

Electric shock

In giving first aid to an electric-shock victim, a caregiver must not touch the victim with bare hands until the source of electricity has been removed safely or the power source shut off. If the victim is not breathing, mouth-to-mouth or mask-to-mouth resuscitation is necessary. If the person also has no pulse, cardiopulmonary resuscitation (CPR) should be administered until professional emergency help arrives. Burns should be rinsed with or immersed in cold water, blotted dry, and kept clean and covered until the victim can be examined by a physician. First aid also includes the prevention of shock, a reduction of blood flow to body tissues that can cause increased anxiety; pale, cool, clammy skin; rapid, weak pulse; possible fainting; or in more serious cases, coma or death.

The controlled delivery of an electric shock, called defibrillation, is used to restart the heart after a heart attack, which may result from accidental electric shock that stops the heart from beating properly. It also is used to restart the heart during open-heart surgery, and for the treatment of some mental illnesses, especially acute forms of depression, through a procedure called electroconvulsive therapy (ECT).

Artificial respiration

Introduction

Artificial Respiration is a process by which air is forced in and out of the lungs of a person by another person or by mechanical means in order to restore normal respiration.

When artificial respiration is used

It is usually employed during suspension of natural respiration caused by disease, such as poliomyelitis or cardiac failure; by electric shock; by an overdose of depressive drugs such as morphine, barbiturates, or alcohol; or by suffocation resulting from drowning, breathing noxious gases, or blockage of the respiratory tract. If the brain is deprived of oxygen for five minutes, it may be permanently damaged; slightly longer periods without oxygen usually result in death. The exception is drowning in very cold water, in which the body's oxygen demand is greatly reduced; people have been revived after being submerged for one-half hour in cold water.

Principles behind exhaled air resuscitation

- The volume of air remaining in the lungs during respiratory failure is maintained by equilibrium between the elasticity of the lungs and the elasticity of the ribs or rib cage. During respiratory collapse, the elasticity of the lungs tends to drive out air from the lungs, whilst the elasticity of the ribs or rib cage draws air into the lungs. There is thus a negative pressure between the outside surface of the lungs and the inside surface of the chest wall. This negative pressure further helps in dilating the capillary network of the lungs and ensuring that at all times there is an adequate circulation of blood in the lungs for gas exchange, for instance, oxygen in and carbon dioxide out.

  In addition, this balance of opposing elastic forces ensures that when the lung is inhaled the circulating blood is not driven out, except during extreme and unnecessary over-inflation. In normal respiration, just as with introduction of air by exhaled air resuscitation, this balance is upset to produce inspiration. Its return to normal gives expiration.

- Expired air does not necessarily contain carbon dioxide;—it is often suggested that expired air is unsuitable because it contains carbon dioxide. This not true. It has been established that fresh air contains approximately 21% oxygen and 79% nitrogen, with an insufficient trace of carbon dioxide. Expired air, though it contains 3% carbon dioxide, still contains 18% oxygen. Thus there is plenty of oxygen in expired air than it has always been thought of.

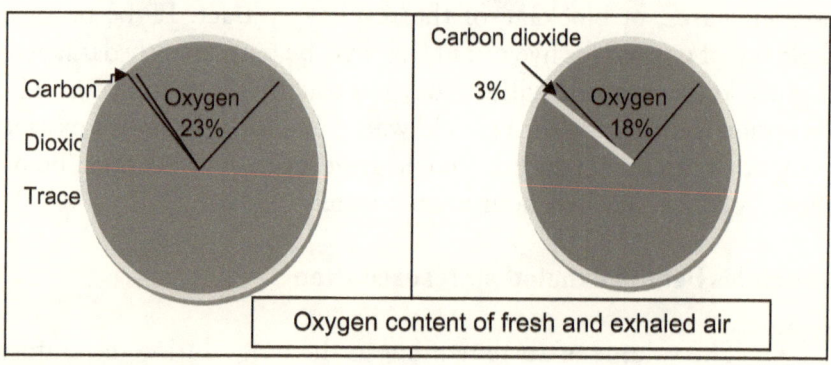

Oxygen content of fresh and exhaled air

Figure 2 Oxygen content of fresh and exhaled air

Human intervention techniques

Because of the danger to the brain of even short periods without oxygen, *artificial respiration or exhale air* resuscitation should always be started immediately. Before commencing exhaled air resuscitation, ensure that

- There is a clear airway from the nose and mouth to the lungs
- Remove any debris, false teeth, water, vomit
- The casualty is on his back
- Casualty's level of consciousness is determined

The **mouth-to-mouth (kiss of life) or mouth-to-nose methods**, shown to be superior to the **back-pressure** and **arm-lift** procedures, is now recommended by the International Red Cross.

Mouth-to-mouth (kiss of life)

- In the *mouth-to-mouth method* the unconscious casualty is laid face up on a firm surface.
- Loosen tight clothing on the chest.
- Then next, quickly remove any debris or food particles from the mouth or throat using fingers, or a napkin or handkerchief or an aspirator as may be available.
- With one hand hold the lower jaw forward with the mouth opened. Make sure your thumb is not placed inside the mouth—it may be bitten during recovery.
- The thumb and forefinger of the other hand keep the nostrils closed whilst the wrist steadies the extended head. (The neck is lifted and the head tilted as far back as possible to prevent the tongue from blocking the air passages, but do it such that the head is not unduly tilted as to cause a constriction of the air passages).
- Take a fairly deep breath of fresh air
- The victim's nose is then pinched shut, and with the reviver's mouth tightly covering that of the victim, the reviver gives four quick, deep breaths.
- If breathing does not resume, proceed to give one breath every five seconds, allowing the air to come out of the victim's lungs between breaths. This is continued until the victim resumes breathing or until trained help arrives.
- If the unconscious person is a baby or small child, cover both the mouth and nose with your mouth, and small puffs of air are breathed out to the victim at the rate of one every three seconds.

Disadvantages of mouth-to-mouth resuscitation

- It is unhygienic

- Gross facial injury or the presence of poisonous material on the victim's face may make the method impossible.
- It is discomforting for both the care giver and casualty

Mouth-to-nose technique

In this technique, air is blown through the nose, during which the resuscitator is simultaneously shifting upwards the lower jaw to prevent the tongue from retracting. The mouth of the casualty is kept closed during inflation by the hand which is supporting the jaw. During exhalation it is an advantage to use this hand to open the victim's mouth.

Advantages of exhaled air resuscitation (EAR)

- *Efficiency*, as EAR gives a more effective inflation of the lungs than other methods
- *Gentleness;* it causes the minimal physical disturbance of the casualty thus improving chances of recovery.
- *Ease of performance*; it is less exhausting, can be maintained over a long period and it is possible for a child to resuscitate an adult.
- *Convenience;* it can be performed on casualties in awkward positions. What is important is that the face is visible and the chest is free to move.
- *Easy to teach others;* it can be practised during emergency health lectures, at the clinical skills laboratory and easy to practise on manikins or models

Sylvester Brosche method

This is one of resuscitation by manual method. The procedure include the following steps

- The patient is placed on his back with a firm cushion or sand mound, between the shoulder blades or the scapulae. This allows the head to fall back opening the airway.

- Kneel comfortably behind the casualty's head facing the feet and grasp the forearms just below the elbows.
- Induce inspiration by drawing the arms upwards, outwards and backwards in a sweeping movement above the head as far as they will go, that is, until they come in contact with the ground
- To allow and aid expiration, sweep the arms back on to the casualty's ribs and continue to squeeze the chest by pressing on the front and sides.
- Repeat the cycle about twelve times every minute. When natural breathing returns the cycle should be continued in step with it until it is established without assistance.

The Heimlich manoeuvre

To restore breathing to a person who is choking, a rescuer gives four quick blows between the victim's shoulder blades with the heel of the hand. If this does not dislodge the obstruction, the rescuer uses the stomach thrust, popularly called the **Heimlich manoeuvre** after its developer, the American physician Henry Jay Heimlich.

Procedure for Heimlich manoeuvre

The rescuer places the side of the fist against the victim's stomach, below the ribs and above the navel. Then, using the other hand, the rescuer thrusts the fist up into the victim's stomach forcefully four times. With children, a rescuer first turns the child head-down and slaps the child's back. In applying the Heimlich manoeuvre to children, the rescuer uses only the first hand, and not the second.

Respirators

Mechanical devices for the administration of artificial respiration include a portable resuscitator used by police and fire departments and the heart-lung machine used to maintain oxygen saturation in the blood during open-heart surgery. Severe breathing difficulties may require help from a mechanical ventilator, which forces air into the lungs by way of a tube inserted into the upper airway through the nose, mouth, or a slit in the trachea. Comatose patients dependent on such a ventilator for a prolonged period may not resume spontaneous breathing. In the much-publicized case of Karen Ann Quinlan in 1976, the New Jersey Supreme Court ruled that a mechanical ventilator could be disconnected under certain conditions so that the comatose patient could "die with dignity." Normal breathing resumed, and the patient lived; nevertheless a precedent was established for the removal of life-support ventilators in the absence of electrical activity in the brain cortex.

Tracheostomy is an emergency operation to introduce into the trachea a tube through an incision made in the anterior surface of the neck. It may also be used in cases of asphyxia due to diphtheria or false croup, or when it is caused by foreign bodies clogging the throat or when the throat is injured

Any tube can be used (the neck of a teapot, a roll or a metal tube) when a special tracheostomy tube is unavailable. The wound rapidly heals after the tube has been removed.

Resuscitation by artificial respiration can prevent the death of a person with water in the lungs if instituted quickly. Because of the constant need of body tissues for oxygen, even a few minutes of suffocation can result in brain damage or death. The exception to this appears in persons who have been submerged in cold water. Some victims have been completely revived, without brain damage, after having been underwater for as long as a half hour. This phenomenon, the so-called diving reflex, has long been observed in sea mammals. Activated when the face is plunged into water below 21° C (70° F), it slows body processes so that oxygen-bearing blood is diverted to the heart and brain.

CHAPTER 9

PRINCIPLES OF MANAGING AND CARING FOR BURNS AND SCALDS

Introduction

A burn is an injury to any part of the body as a result of exposure to dry heat, such as fire or a piece of hot material, for instance, flames or glowing coal. A scald on the other hand is caused by the effects of moist heat, for examples, boiling water, steam, hot tar or fat.

When tissue is burned or scalded, there is damage to the capillaries with escape of fluid into the tissue causing oedema. The oedema exerts pressure on blood vessels resulting in further death of tissue. In a burn or scald, even when the source of heat is removed, the heat retained in the tissues continues to cause pain and damage. Fluid loss as a result of burns or scalds causes shock, infection and if not properly cared for can lead to septicaemia.

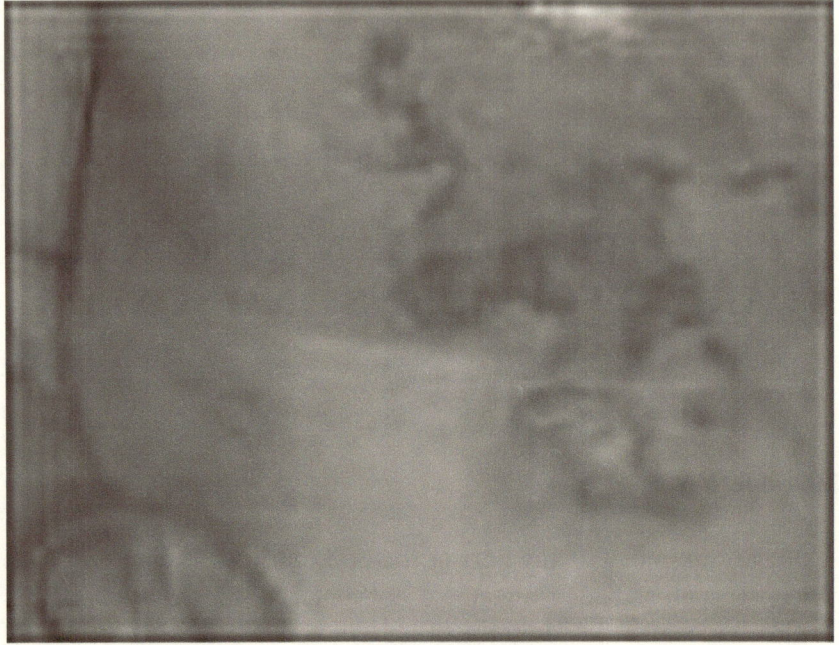

Figure 4 Scalded skin

Degrees of severity of burns or scalds

First degree (erythema)—is when skin turns red, swells and tender to touch. It is the mildest degree of burn.

Characteristics

- Skin inflammation (which disappears after 3-6 days)
- Pigmentation and peeling of skin later at the site of affection
- pain

Second degree (formation of blisters)

Characteristics

- more pronounced inflammatory reaction
- severe pain
- bright red lesions

- blisters filled with a transparent or slightly cloudy fluid form
- intact deeper skin layers
- if not infected, skin restores without scaring within 7 days, with full recovery in 10-15 days
- if infected, wound heals by second intention

Third degree—when all the skin layers suffer necrosis

Characteristics

- dense scab on injured and necrotic tissues
- wound heals by second intention
- granulation forms at the site of affection
- replaced by connective tissue and a star-like scar

Fourth degree (charring occurs at this stage)—often caused by electrical arcs or molten metal. It is the gravest form of burn.

Characteristics

- Damage to skin, muscles, tendon and bones
- Heals slowly
- Often needs skin grafting to cover the burnt surface

Depth of Burns

Burns can be categorised as follows:

Superficial burns

These involve only the outer layer of the skin, and are characterised by redness, swelling and tenderness. Typical examples are mild sunburn, or a scald produced by a splash of hot tea or coffee. Superficial burns usually heal well if prompt first aid is given, and do not require medical treatment unless extensive.

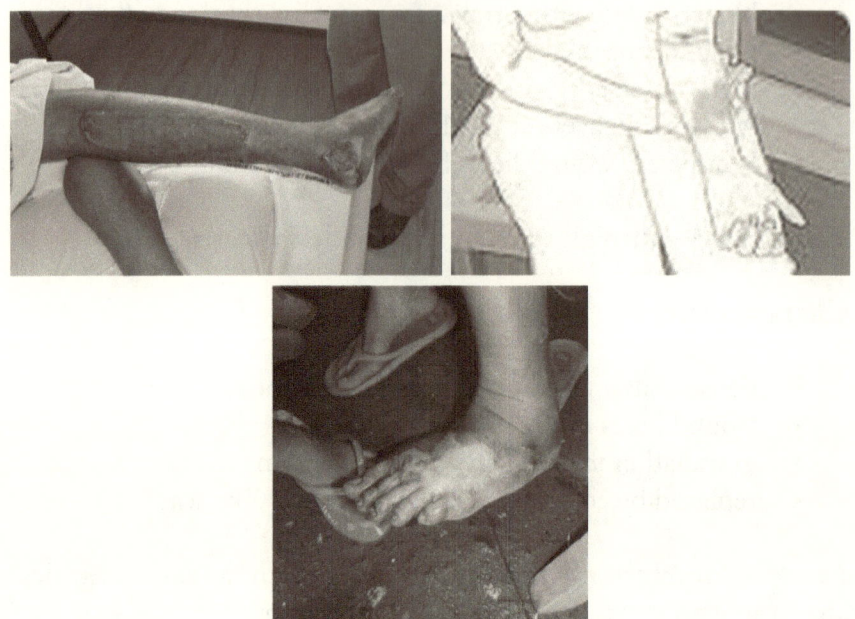

Figure 5 superficial burns

Partial-thickness burns

These damage a 'partial thickness' of the skin, and require medical treatment. The skin looks raw, and blisters form. These burns usually heal well, but can be serious, if extensive. In adults, partial-thickness burns affecting more than 50% of the body's surface can be fatal. This percentage is less in children and the elderly.

Full-thickness Burns

These damage all layers of the skin. Damage may extend beyond the skin to affect nerves, muscle and fat. The skin may look pale, waxy, and sometimes charred. Full-thickness burns of any size always require immediate medical attention, and usually require specialist treatment.

Extent of Burns

The area of a burn gives an approximate indication of the degree of shock that will develop and, in conjunction with depth, can be used as a guide to the required level of treatment. The Wallace's 'Rules of Nine' is a guide used to calculate the extent of a burn as a percentage of the body's total surface area, and to assess what level of medical attention is required.

The percentage distribution in the Rule of Nine is as follows

Face -	**9%**
Front (trunk and chest areas) -	**18%**
Back (covering entire back) -	**18%**
The Private parts region -	**1%**
The lower limbs (thigh and leg areas) -	**18% each**
The upper limbs (arm and hand areas) -	**9% each**

In a healthy adult:

Any *partial-thickness burn* of 1% or more (an area approximating to that of the casualty's hand) must be seen by a medical practitioner.

A partial-thickness burn of 9% or more will cause shock to develop, and the casualty will require hospital treatment.

A *full-thickness burn* of any size requires hospital treatment.

Severe Burns and Scalds

The priority is to cool the injury; the longer the burning goes unchecked, the more severely the casualty will be injured. Resuscitate the casualty only when cooling is underway. All severe burns carry the danger of shock.

Factors influencing the effects of burns and scalds on the body

- The degree of severity

- The surface area involved
- The site of the body affected
- The age of the victim

Complications

- Shock—as a result of emotional and physical impact of the extensive nature of the burns or scalds, especially when exposed to cold, severe pain, absorption of toxic substances from the burnt tissues.
- Septic absorption—

Assessing a Burn

There are a number of factors to consider when assessing the severity of a burn and the method of treatment, including the

- Cause of the burn, whether the airway is involved, the depth of the burn, and its extent. If the airway is involved, depending on the severity, the casualty might need urgent tracheostomy to maintain an airway. The cause of the burn may also signal any other possible complications.

Figure 6 Exposure of children to burns and scalds

- The extent of the burn will indicate whether shock is likely to develop, as tissue fluid (serum) leaks from the burned area and is replenished by fluids from the circulatory system. The greater the extent of the burn, the more severe the shock will be.
- Burns also carry a serious risk of infection, which increases according to the size and depth of the burn. The body's natural barrier, the skin, is destroyed by burning, leaving it exposed to germs.

Treatment

Principles of treatment

- To get rid of residual heat
- To control shock
- To prevent infection

First aid for most burns is cool water applied soon after the burn. Application of home remedies should be avoided. Burns of 15 percent of the body surface or less are usually treated in hospital emergency rooms by removing dead tissue (debridement), dressing with antibiotic cream (often silver sulfadiazine), and administering oral pain medication.

Burns of 15 to 25 percent often require hospitalization to provide intravenous fluids and avoid complications. Burns of more than 25 percent are usually treated in specialized burn centres where aggressive surgical management is directed toward early skin grafting and avoidance of such complications as dehydration, pneumonia, kidney failure, and infection.

Pain control with intravenous narcotics is frequently required. The markedly increased metabolic rate of severely burned patients requires high-protein nutritional supplements given by mouth and intravenously. Extensive scarring of deep burns may cause disfigurement and limitation of joint motion. Plastic surgery is often required to reduce the effects of the scars. Psychological problems often result from scarring.

Prompt treatment of burns and scalds may help to limit damage and alleviate pain. Treatment is outlined below:

Severe Burns

- Start cooling the burn immediately under running water for at least 10 minutes
- Dial for an ambulance
- Lay the casualty down, protecting the burned area from contact with the ground, if possible and make the casualty as comfortable as possible
- Continue to pour copious amounts of cold water over the burn for at least ten (10) minutes or until the pain is relieved
- You should remove all jewellery or clothing from the affected area, unless it is sticking to the skin. However ensure that you are wearing disposable gloves before doing this.
- Put a clean, non-fluffy material over the burn to protect from infection. Cloth, a clean plastic bag or cling film all make good dressings.
- Treat for shock

Minor Burns

- For minor burns, run cold water over the affected area for a minimum of 10 minutes or until the pain eases.
- Remove any jewellery etc. and cover the burn as detailed above.
- If a minor burn is larger than a postage stamp it requires medical attention. All deep burns of any size require urgent hospital treatment.

Clothing on fire

- Stop the casualty panicking or running—any movement or breeze will fan the flames.

- Drop the casualty to the ground.
- If you can, wrap a coat, blanket or curtain (not the nylon or cellular type), rug or other heavy-duty fabric tightly around the casualty. The best fabric is wool.
- Roll the casualty along the ground until the flames have been smothered.

On ALL burns DO NOT:

- Apply creams, lotions, ointments or fats to the injured area
- Use adhesive dressings
- Break blisters or interfere with the injured area

Burns to the Mouth and Throat

Burns to the face, and burns in the mouth or throat are very dangerous, as they cause rapid swelling and inflammation of the air passages. The swelling will rapidly block the airway, giving rise to a serious risk of suffocation. Immediate and highly specialised medical assistance is required.

Treatment of Burns to the Mouth and Throat

Contact the emergency service. Report suspected burns to the airway.

Take any steps to improve the casualty's air supply, e.g., loosening clothing around the neck. Give the casualty oxygen if you are trained to do so.

If the casualty becomes unconscious, place in the recovery position, and be prepared to resuscitate.

Minor Burns and Scalds

Minor burns and scalds are usually the result of domestic accidents. Prompt first aid will generally enable them to heal naturally and well, but the advice of a medical practitioner should be sought if there is doubt as to the severity of the injury.

Treatment of Minor Burns and Scalds

- DO NOT use adhesive dressings.
- DO NOT break blisters, or interfere with the injured area.
- DO NOT apply lotions, ointments, creams, or fats to the injured area.
- Cool the injured part with copious amounts of cold water for about 10 minutes to stop the burning and relieve the pain. If water is unavailable, any cold, harmless liquid such as milk or canned drinks will suffice.
- Gently remove any jewellery, watches, or constricting clothing from the injured area before it starts to swell.
- Cover the injury with a sterile dressing, or any clean, non-fluffy material to protect from infection. A clean plastic bag or kitchen film may be used.

CHAPTER 10

PRINCIPLES OF MANAGING FRACTURES

||||||||||||||||||||||||||||||||||||||||||||||||||||||||||||||||||||||||||||||||||||||||||||||||||||||||||||||||||||

Fracture is a break or crack in the continuity of the bone or in ossified cartilage. Simple, or closed, fractures are not visible on the surface. Compound, or open, fractures involve a rupturing of the skin, often exposing the bone. Single and multiple fractures refer to the number of breaks in the same bone. Fractures are complete if the break is total or incomplete (greenstick) if the fracture occurs only part of the distance across a bone shaft, with bending or crushing of the bone. Incomplete fractures are found mostly in young children, whose bones are resilient. Heavy impact causes most fractures but the simple activity of throwing a ball could cause a break.

Causes

- Sudden violent pressure against bone
- Pathological

Symptoms

Common symptoms of a fracture are

- Severe local pain and increases with movement of the limb or weight on the area of injury,
- Tenderness and crepitations when injured site is palpated. Palpation is unnecessary in open wounds where bone fragments can easily be seen.

- Swelling with some degree of deformity (shortening of the limb). X rays are the only accurate means of detecting and defining the type of fracture.
- Functional disorders (inability to move the limb)
- Abnormal mobility of the limb where there should be none

Infection in compound fractures is treated with antiseptics and antibiotics. If the broken segments lie adjacent to each other, *stretching or traction* to overcome the pull of powerful muscles may be used to achieve realignment, although external manipulation may sometimes bind the segments.

This is called *reduction*. If proper alignment cannot be achieved in this way, an operation is usually performed, and the fragments are joined with screws, nuts, nails, wires or metal plates. This is *open reduction*. Once aligned, segments are secured externally with a plaster cast or splint to immobilize the fracture and to speed healing.

When ribs are fractured, the chest is often strapped or taped to reduce pain from the motion of breathing. While healing, the body creates new tissue to join the broken segments. Bone producing cells invade the granulation tissue and form a bone-like material called *callus*. This begins to appear about 10 days after a fracture and grows in two layers; first, the outer callus holds the fragments together from outside, and the inner callus grows within the bone itself. Minerals in the tissue harden to form solid new bone structure. When this has occurred, the inner and outer layers of callus gradually disappear.

The duration of repair of a broken bone depends on the age of the victim, the size of the bone and the variety of the fracture. Thus bone healing occurs much quickly in childhood, particularly in small bones, such as the ribs and clavicle or the collar bone.

Classification of fractures

Traumatic - this kind of fractures can be closed (the skin remains intact) or open (the skin is damaged) and often occur following an impact on the bone

Pathological - often induced by diseases (tuberculosis, osteomyelitis, tumours) which develop in the bone mass and lead to its gradual destruction. At some stage, the slightest weight can cause the bone to fracture.

Common Fractures

In a *greenstick* fracture, the bone does not break all of the way through. A *compression fracture* results from compression or flattening of the bone. *Wedged fractures* occur when fragments of broken bones are driven into the other. Fractures are called *simple,* or *closed,* when the bone breaks but the skin does not. A *compound,* or *open,* fracture is when the broken bone tears through the skin, introducing the dangerous possibility of infection and can lead to the development of osteomyelitis.

Open fractures make reunion of the bone fragments more difficult. The area around a break swells and discolours, but some fractures can be detected only by X-ray. The weakened bones of the elderly are especially susceptible to fractures. Fractures can also be *complete* or *incomplete.* Complete fractures are distinguishable by the disruption of some part of the bone diameter, most commonly in the form of a longitudinal fissure.

Treatment of fractures in emergency situation

- Immobilise bones at the site of fracture
- Control bleeding by covering with a temporary dressing and stabilised with a bandage
- Control and prevent shock
- Fixation-position of the fracture with bandages or splints
- Prompt removal of victim to a medical establishment
- Observe the victim as he is being treated or transported to hospital

HEAD, EYE, AND NOSE INJURIES

Head

- Injuries to the head may involve the scalp, skull, or brain.
- If the victim has a head wound, do not apply pressure to it, as this may damage the brain.
- Keep victim's airway clear from obstructions, such as vomitus, which is common in cases of head injury.
- If the victim has a seizure, a sudden spasm of the body, the head must be protected with cushions to prevent further injury.
- All individuals with head injuries should be evaluated by a physician.

Eye

First Aid Treatment for Eye Injuries

THE EYES ARE ONE OF THE MOST DELICATE PARTS OF THE HUMAN BODY, SO EYE INJURIES SHOULD BE TREATED PROMPTLY AND ACCURATELY.

A tiny speck of sand in your eye can feel like a razor blade. So a more serious injury such as a cut on the cornea, a black eye, or a piece of foreign object in your eye is a problem that should always be checked out by a physician. Try to get an appointment to see your own family physician or regular eye doctor if you can, but if not, an urgent care centre or emergency room is a good second choice. But for serious injuries, you should never, ever try to treat the problem on your own.

Minor irritations, such as flecks of dirt, rogue eyelashes, or wayward bugs, can easily get into your eye and cause irritation. Tiny things that get in your eye usually feel much bigger, and are much more irritating than they would feel on any other part of your body. A small foreign object can also cause redness, a burning sensation when you blink, or sensitivity to light. These symptoms almost always make you want to rub your eye—but don't! Rubbing an irritated eye can make things worse or even cause permanent damage.

Every time you get something in your eye, it needs to be removed. But emergency medical attention is of particular importance if your vision is affected. If your vision is blurry or you see light specks, waves, or darkness in your field of vision, then you need to see a doctor immediately.

If a person has a foreign object in the eye, you may be able to help him by flushing the eye with cool, clean water. If you are not near a sink, you can use a glass, pitcher, or eye dropper. If you can see the object and it is on the eyeball surface, flushing should remove it.

If you cannot see the object on the surface of the eyeball, then it may be stuck beneath the eyelid. Have the person lie down beneath a good light, or position a flashlight so it shines on the eye but leaves both of your hands free. Have the person look upward, and then gently pull downward on their lower eyelid. If you see a particle on the inside of the eyelid or on the lower part of the eyeball, then you can flush it out with water in an eyedropper, or touch a moistened cotton swab or gauze strip gently to the eye so that it adheres and can be removed.

If you don't see anything on the lower eyelid, then check the upper eyelid. An easy way to lift the upper eyelid is to take a cotton swab, place it on top of the eyelid, and gently curl the eyelashes and upper eyelid over the swab. Take care not to pull or strain the eyelid while pulling. If you can see the object on the inside of the eyelid or on the surface of the eyeball, try to flush it out with water while you hold the curled swab in place. If flushing doesn't work, you can gently touch the speck with a moist cotton swab to see if the cotton can pick it up.

If a foreign object in the eye does not flush out with water, then you should cover BOTH eyes with gauze gently, and seek help immediately. It may seem odd to cover both eyes if only one is affected, but the reason makes perfect sense. Your eyes do not work independently; they operate in conjunction with each other. So if you leave one eye uncovered, then covering the other eye does no good because it will still be moving in concert with the uncovered eye. If you cover both eyes, you are essentially keeping them closed, and thereby preventing movement. Keeping the eyes as still as possible not only prevents further irritation, it also helps to calm and soothe the nerves that are certainly on edge because of the pain and irritation in the eyes. Also, being sure that both eyes are covered to reduce movement will minimize damages if there is an object embedded in the eyeball.

Your eyes are delicate and usually irreplaceable, so you need to treat them with tender loving care. It is common to get something in your eye now and then, but if you don't know how to safely handle the problem, your eyesight may suffer for the rest of your life. Try these simple tips:

- Medical attention should be sought for all eye injuries as well.
- In the case of foreign material in the eye, especially *caustic* substances, or those that can burn, corrode, or dissolve tissues, the eye should be flushed immediately with a cool, sterile saline solution, if available, or plain tap water for 15 to 30 minutes.
- Do not attempt to remove embedded objects from the eye.

Nose

The most common injuries to the nose involve nosebleeds, objects lodged in the nasal passages, and broken nasal bones.

- The victim of a simple nosebleed should sit down, lean forward, and gently pinch together the soft part of the nose for 15 minutes.
- A cold compress can also be placed on the bridge of the nose.
- If material lodged in the nose cannot be forced out by gently blowing the nose, the victim should request medical help.
- In the case of a broken nose, apply a cold compress to the bridge of the nose and seek medical attention.

Cold injuries

- Exposure to cold can lead to **hypothermia,** a condition in which the body's internal temperature drops below normal. The first-aid provider should seek medical help first, if possible, and re-warm a hypothermic individual by whatever means available, including body warmth. If the victim is alert, warm, sweet fluids may be given. If the victim is breathing at a rate less than six breaths per minute, mouth-to-mouth or mask-to-mouth resuscitation can be started. *CPR should not be performed because a hypothermic person may have a heartbeat even when the pulse is undetectable and any CPR may cause cardiac arrest.*
- **Frostbite** is is a condition in which the skin freezes, initially causing pain and redness in the affected area, which may develop into numbness and whiteness. The first-aid provider should rewarm frozen areas (usually extremities) of the victim's body slowly by using skin to skin contact, immersing frozen part in warm, not hot, water, or using warm compresses. Avoid massaging the affected area, which may cause tissue damage. The first-aid provider should not thaw frozen areas that may refreeze

CHAPTER 11

GENERAL PRINCIPLES AND CAUSES OF UNCONSCIOUSNESS

Unconsciousness is a state in which a person has reduced awareness of his or her surroundings, is without deliberate thoughts, and is less than normally responsive to stimuli such as light and sound.

Unconsciousness is divided into two, **normal unconsciousness** such as sleep, dreams; and **abnormal unconsciousness** which occurs as a result of some pathological effects, viz. Epilepsy, syncope, infantile convulsions, apoplexy, diabetes, uraemia, etc. For the purposes of this course, we shall discuss some selected abnormal unconsciousness only as seen in the proceeding paragraphs below.

Epilepsy

Epilepsy, also called seizure disorder, is a chronic brain disorder that briefly interrupts the normal electrical activity of the brain to cause seizures, characterized by a variety of symptoms including uncontrolled movements of the body, disorientation or confusion, sudden fear, or loss of consciousness. Epilepsy may result from a head injury, stroke, brain tumour, lead poisoning, genetic conditions, or severe infections like meningitis or encephalitis. In over 70 percent of cases no cause for epilepsy is identified. Some 40 to 50 million people suffer from epilepsy worldwide and the majority of cases are in developing countries. According to the World Health Organization (WHO), an estimated 2 million new cases are diagnosed each year globally.

Types of seizures

Epileptic seizures vary in intensity and symptoms depending on what part of the brain is involved. In partial seizures, the most common form of seizure in adults, only one area of the brain is involved. Partial seizures are classified as simple partial, complex partial (also known as psychomotor), and absence (also known as **myoclonic or petit mal**) seizures.

People who have simple partial seizures may experience unusual sensations such as uncontrollable jerky motions of a body part, sight or hearing impairment, sudden sweating or flushing, nausea, and feelings of fear.

Complex partial seizures, also called temporal lobe epilepsy, last for only one or two minutes. The individual may appear to be in a trance and moves randomly with no control over body movements. The individual's activity does not cease during the seizure, but behaviour is random and totally unrelated to the individual's surroundings. This form of seizure may be preceded by an *aura* (a warning sensation characterized by feelings of fear, abdominal discomfort, dizziness, or strange odours and sensations).

Absence seizures, are known to be rare in adults, and are characterised by a sudden, momentary loss or impairment of consciousness. Overt symptoms are often as slight as an upward staring of the eyes, a staggering gait, or a twitching of the facial muscles. No aura occurs and the person often resumes activity without realizing that the seizure has occurred.

In a second type of epilepsy, known as **generalised seizure**, tonic clonic, grand mal, or convulsion, the whole brain is involved. This type of seizure is often signalled by an involuntary scream, caused by contraction of the muscles that control breathing. As loss of consciousness sets in, the entire body is gripped by a jerking muscular contraction. The face reddens, breathing stops, and the back arches. Subsequently, alternate contractions and relaxations of the muscles throw the body into sometimes violent agitation such that the person may be subject to serious injury. After the convulsion subsides, the person is exhausted and may sleep heavily. Confusion, nausea, and sore muscles are often experienced upon awakening, and the individual may have no memory of the seizure. Attacks occur at varying intervals, in some people as seldom as once a year and in others as frequently as several times a day. About 8 percent of those subject to generalized seizures may have *status epilepticus,* in which seizures occur successively with no intervening periods of consciousness. These attacks may be fatal unless treated promptly with the drug diazepam.

Diagnosis

In persons suffering from epilepsy, the *brain waves,* electrical activity in the part of the brain called the cerebral cortex; have a characteristically abnormal rhythm produced by excessive electrical discharges in the nerve cells. Because these wave patterns differ markedly according to their specific source, a recording of the brain waves, known as an electroencephalogram (EEG) is important in the diagnosis and study of the disorder. Diagnosis also requires a thorough medical history describing seizure characteristics and frequency.

Treatment

There is no cure for epilepsy but symptoms of the disorder may be treated with drugs, surgery, or a special diet. Drug therapy is the most common treatment—seizures can be prevented or their frequency lessened in 80 to 85 percent of cases by drugs known as anticonvulsants or anti-epileptics. Surgery is used when drug treatments fail and the brain tissue causing the seizures is confined to one area and can safely be removed.

A special high-fat diet known as a ketogenic diet produces a chemical condition in the body called *ketosis* that helps prevent seizures in young children. Like any medical condition, epilepsy is affected by general health. Regular exercise, plenty of rest, and efforts to reduce stress can all have a positive effect on a person with a seizure disorder.

First aid for generalised seizures involves protecting the individual by

- Clearing the area of sharp or hard objects, providing soft cushioning for the head, such as a pillow or folded jacket and, if necessary, turning the individual on the side to keep his or her airway clear. The individual having a seizure should not be restrained and the mouth should not be forced open—it is not true that a person having a seizure can swallow the tongue. If the individual having the seizure is known to have epilepsy or is wearing epilepsy identification jewellery, an ambulance should only be called if the seizure lasts longer than five minutes, another seizure closely follows the first, or the person cannot be awakened after the jerking movements subside.

Fainting (Syncope)

Fainting is a sudden dizziness or weakness accompanied by brief loss of consciousness, associated with insufficient oxygen in the brain.

Causes

- Disturbance in blood circulation due to fatigue, pain, shock, abnormal blood pressure, arterial blockage, or heart failure.
- The person fainting should be placed in a position that will quickly bring blood to the brain, and other aid to blood circulation should be instituted. The cause of fainting should be quickly determined so that further appropriate action can be taken.

Apoplexy (Stroke, apoplectic seizure, coma)

Apoplexy is state of unconsciousness in which a person is unresponsive to external stimuli. It occurs suddenly in elderly people and due to haemorrhage into the substance of the brain, or to clotting of blood in the cerebral blood vessels. The disease is known technically as cerebral haemorrhage or thrombosis, and embolism according to whether it is due to bleeding or clotting respectively in the brain.

In the deepest coma, spontaneous respiration ceases, and a mechanical respirator must be used. Coma may last for a few days or, in rare cases, for years, usually progressing after the first month to a persistent **vegetative state**. Coma in which electrical activity can no longer be detected in the brain is called **brain-death syndrome**.

Causes

- stout men with high blood pressures are more susceptible
- Arteriosclerosis (hardening of the arteries)
- Decreased metabolic activity in the brain, which may be caused by cerebral haemorrhage, inflammation of the brain due to meningitis or encephalitis, drug overdose, oxygen deprivation (as in cardiac arrest), or
- Abnormal metabolism. Among the conditions that can cause metabolism abnormalities are diabetic ketoacidosis, in which the blood is too acidic; a high blood level of ammonia, which often follows liver damage caused by alcohol; or uraemia, in which damaged kidneys cannot process the toxic waste products of metabolism. Most people recover. Some develop "respirator brain," a poorly understood condition in which the body cannot resume breathing on its own.

Signs and symptoms

- Onset—persistent headache, shortness of breath, nose-bleeding and giddiness
- Severe pain in the head
- Fainting
- Slow and strong pulse
- Characteristic breathing, accompanied by snoring and puffing out of the cheeks with expiration
- Unequal pupils and unresponsive to light
- Head and eyes often turned towards the affected side of the brain.

Treatment

- Avoid undue examination of the casualty
- Do not move casualty more than it is absolutely necessary
- Position victim's head and shoulders slightly raised on a pillow, with the head turned towards the affected side.
- Care of the mouth—remove any false teeth, wipe away saliva with small wisps of cotton wool wrapped round the little finger or any suitable holder. Caring for the mouth is important because elderly people often have septic teeth, and if the mouth is allowed to fill up with saliva and thereby collect germs from the unhealthy gums, there is always the risk that some of the saliva may be sucked down into the lungs and cause broncho-pneumonia, common and often fatal complication.
- Keep casualty warm by covering with a blanket or any suitable wraps placed under him and tucked well in at the sides and feet
- *Masterly inactivity* is the most important treatment to adopt after the above procedures have been undertaken. It comprises doing absolutely nothing beyond watching the client and appreciating the fact that additional treatments, such as trying to give stimulants or medicines, will do more harm than good. Masterly inactivity is one of the most difficult treatments to apply in medical work and first aid.
- *Medical advice* especially where expert care in special hospital is needed.

Infantile convulsions

Fits occurring in infants and children are known as infantile convulsions. These are common up to the age of 18 months. However, convulsions also occur in older children.

Causes

- Reflex causes—any irritation that can provoke a fit. Examples include, constipation, indigestion, ear-ache, worms infestation, teething or severe malaria.
- Infectious diseases—convulsions in infants and children often occur at the onset of an infectious disease, such as scarlet fever, measles, chicken pox, influenza. In a similar pattern, just like children adults suffer rigor instead of convulsions.
- Malaria, enteric fevers, urinary tract infections (UTI), meningo-encephalitis

Signs and Symptoms

- General nervous irritability
- Breathe—holding—the infant develops irregularity in breathing, or may stop breathing for a second or two.
- Rigidity—he throws his head back and becomes stiff all over
- Altered colour—the face is pallor, bluish of face or limbs
- Twitching
- Squinting and frothing at the month occasionally

Treatment

- Reassure parents to calm down
- Place the child on a warm bed or couch
- Apply the principles involved in treating unconsciousness
- Send for medical assistance to determine cause of fit and further management

CHAPTER 12

POISONS AND MANAGEMENT IN POISONING

‖‖

Introduction

A poison is any substance which is liable to have a harmful action on the human body, injuring health or destroying life. Put differently, poison is any substance that produces disease conditions, tissue injury, or otherwise interrupts natural life processes when in contact with or absorbed into the body. Poison may enter the body in three ways

- Orally
- Inhalation
- Through the skin
- The axial route

Most poisons taken in sufficient quantity are lethal. A poisonous substance may originate as a mineral, vegetable, or an animal and it may assume the form of a solid, liquid, or gas. A poison, depending on the type, may attack the surface of the body or, more seriously, internal organs or the central nervous system.

Poisoning may be

- Accidental
- Suicidal
- Homicidal

The presence of suspicious bottles, pills or leaking gas could suggest poisoning. If suicide or homicide is suspected, do not remove or destroy any likely evidence in case of forensic investigation. A priority however would be how to save the life of victim as soon as possible.

Types of poisons

Poisons in humans are usually classified according to their effects as corrosives, irritants, or narcotics; the last named are also known as systemic or nerve poisons.

- Corrosives include strong acids or alkalis that cause local tissue destruction, externally or internally; that is, they "burn" the skin or the lining of the stomach. Vomiting occurs immediately, and the vomitus is intermixed with blood. Common or so-called household corrosive poisons include hydrochloric acid, carbolic acid, bichloride of mercury, and ammonia.
- Irritants such as arsenic, mercury, iodine, and laxatives act directly on the mucous membrane, causing gastrointestinal irritation or inflammation accompanied by pain and vomiting; diluted corrosive poisons also have these effects. Irritants include cumulative poisons, those substances that can be absorbed gradually without apparent harm until they suddenly take effect.
- Narcotic poisons (systemic or nerve poisons) act upon the central nervous system or upon important organs such as the heart, liver, lungs, or kidneys until they affect the respiratory and circulatory systems. These poisons can cause coma, convulsions, or delirium. Narcotic poisons include alcohol, opium and its derivatives, belladonna, turpentine, potassium cyanide, chloroform, and strychnine. Also included in this category is one of the most dangerous poisons known, botulin toxin, and a potent bacterial toxin that is the cause of acute food poisoning
- Blood poisoning, also bacterial in nature, is a condition that occurs when virulent micro-organisms invade the bloodstream through a wound or an infection. Symptoms

include chills, fever, prostration, and often infections or secondary abscesses in various organs. Most poison gases also affect the bloodstream. Because these gases restrict the body's ability to absorb oxygen, they are often considered in a separate category called asphyxiants, to which group ordinary carbon monoxide belongs. Gas poisons, however, may also be corrosives or irritants

- Commonly used drugs—About 50 percent of all human poisoning cases involve commonly used drugs or household products such as aspirin, barbiturates, insecticides, and cosmetics. Because barbiturates are easily available, toxic effects resulting from their misuse are not infrequent. Acute poisoning may result from over dosage or interaction with other drugs, especially alcohol. The victim of acute barbiturate poisoning may become agitated and nauseated, or may pass into a deep sleep marked by increasingly shallow respiration. Coma and heart failure may follow. Chronic barbiturate poisoning, caused by prolonged use of the drugs, is usually marked by gastrointestinal irritation, loss of appetite, and anaemia. In advanced stages of chronic barbiturate poisoning the victim may show mental confusion.

Treatment

Various treatments may counteract the effect of a poison. The immediate requirement is to ensure that respiration and circulation are maintained; and then decide what type of poison has been used and act as follows:

- Use of a dilution such as, the ingestion of large quantities of water or milk.
- Use of an emetic—an emetic is a substance that induces vomiting and rids the stomach of certain poisons. An emetic may act locally, as on the gastric nerves, or systematically on the part of the brain that causes the vomiting. Household emetics, which act locally, include a tablespoon of salt dissolved in warm water or two tablespoons of mustard dissolved in a pint of water. *Emetics must not be given to a person who has*

swallowed a corrosive poison. An emetic can be given in cases such as in aspirin poisoning, barbiturates poisoning, opium and its derivatives (at least in the early stages of poisoning), convulsants (strychnine, prussic acid and cyanides) and large doses of alcohol.

- Use of an antidote, unlike an emetic, is a remedy that counteracts the effects of a poison chemically, although it may result indirectly in vomiting. An antidote may work against a poison by neutralizing it, rendering it insoluble, absorbing it, isolating it, or producing an opposite physiological effect generally. For instance, in acid or other corrosives poisoning.
- Complex forming absorbent agents usage

In any instance of poisoning, it is imperative that remedial treatment be started immediately.

Food poisoning

Food-borne Illness or Food Poisoning is any illness associated with eating food contaminated by disease-causing bacteria, viruses, or parasites; natural toxins in plants and animals, such as mushrooms and shellfish; or harmful chemical agents such as insecticides and heavy metals. The symptoms of food-borne illness develop within a period of several hours to two days after eating contaminated food and usually include

- nausea,
- abdominal and stomach cramps,
- vomiting, and
- Diarrhoea.
- *Dehydration* (excessive fluid loss from the body) may develop, leading to thirst, dizziness, or fainting.

Although most cases of food-borne illness are generally mild and last for only a few hours, some forms may be life-threatening. A physician should always be consulted, particularly in cases involving infants and children, pregnant women, the elderly, and people with compromised immune systems, who are generally more susceptible to dehydration and other complications. In addition, suspect samples from a recent meal should be saved to help medical personnel determine the source of the illness.

Food-borne illness is commonly caused by certain bacteria or their toxins, which are poisonous proteins produced by these bacteria. One toxin-producing bacterium is *Staphylococcus*, which occurs almost everywhere and grows readily in foods stored at room temperature, especially processed meat and fish, milk, and cream-filled foods.

Escherichia coli, commonly known as *E. coli*, is a species of bacteria normally present in human intestines. A recently recognized strain, *E. coli* 0157:H7 produces high levels of toxins that can cause kidney damage, as well as septicaemia, or blood poisoning. Illness from *E. coli* may develop from consuming undercooked beef, unpasteurized milk, or from handling food without washing hands after changing diapers.

Symptoms can include

- Diarrhoea,
- chills,
- headaches, and
- High fever, and in some cases the infection can lead to death, even with medical intervention.

Botulism is an often fatal disease that results from eating improperly canned foods contaminated with toxins released by the bacteria *Clostridium botulinum*. Although commercial canning methods have made the occurrence of this disease relatively rare, home-canning practices in which food and the container are not thoroughly heated may result in botulism.

In some cases, food-borne illness is caused not by toxins but by rapidly growing colonies of the bacteria themselves. Among the most common of these harmful bacteria is *Salmonella enteritidis*, which is most often spread through poultry, eggs, and egg products such as mayonnaise. Eating undercooked poultry, using cooking utensils and cutting boards used for the preparation of raw poultry without properly cleaning them, and eating eggs or egg products that were not properly refrigerated are the primary causes of infection with *Salmonella enteritidis.*

Recently, scientists have discovered additional bacterial pathogens that can cause food-borne illness. ***Listeria monocytogenes****,* which can cause septicaemia, meningitis, and stillbirth, kills up to one-third of the people infected and most often results from unsanitary commercial processing of dairy, poultry, and meat products—including pizza toppings. **Campylobacter jejuni** is now the most common bacterial cause of diarrhoea in industrialized countries. Caused by contaminated raw foods, *Campylobacter jejuni* is the most prevalent pathogen in poultry, and in more serious cases can result in arthritis, septicaemia, meningitis, inflammation of the heart and other organs, and Guillain-Barré syndrome (paralysis).

Chief among the viruses that cause food-borne illness is **hepatitis A,** which is excreted in the faeces of infected individuals and re-enters the food chain through unsanitary methods of food preparation. Water, salads, shellfish, and milk products are common sources of contamination. Hepatitis A infection causes a mild illness with symptoms that include fever, fatigue, nausea, loss of appetite, soreness in the abdominal area, and *jaundice* (accumulation of the pigment bilirubin in the blood, which causes yellowish discoloration of the skin and whites of the eyes). Other viruses include rotaviruses, which cause acute upset of the digestive system (gastroenteritis); and the Norwalk virus family, which also causes gastroenteritis.

Parasites, such as the *Trichinella spiralis* worm, can cause food-borne illness when humans eat meat from an infected host animal. In addition to initial gastrointestinal symptoms, parasitic infections can cause permanent damage to the eyes, heart, and other organs. Commercial food processing has greatly reduced the incidence of food-borne parasites; however, foods prepared under unsanitary processing conditions or at home—notably, cured or smoked meats—may harbour these pathogens.

Some mushrooms contain natural toxins that may be poisonous when eaten. Poisonous fungi of the *Amanita* species are the source of most mushroom poisonings, causing symptoms that range from mild stomach upset to death as a result of severe liver damage. Certain shellfish, such as mussels, clams, oysters, and scallops, become poisonous in the human digestive system if the shellfish have fed on toxin-producing planktonic algae. Similar algae-produced toxins may work their way up the food chain, becoming concentrated in large, predatory fish such as grouper or barracuda. Ciguatera fish poisoning occurs when people consume these fish. Sushi, when prepared from puffer fish flesh, may contain high levels of toxins if the fish's internal organs—where toxins concentrate—are not carefully removed and discarded.

Prevention methods

Most food-borne illness may be prevented by

- Observing strict sanitary measures in preparing and storing food,
- Serving food soon after preparation, and quickly placing leftovers under refrigeration.
- Food processing techniques, from canning to irradiation (exposure of food to low levels of ionizing radiation), can protect consumers from food-borne illness as well.

Despite these safety features, public health officials and scientists warn that there is a growing risk of food-borne illness in world wide. **Factors contributing to this risk** include an increase in imported produce, often from developing countries with varying safety standards; a growing trend toward eating out, which leaves consumers more vulnerable to pathogens introduced by food handlers; and newly resistant and adaptive strains of bacteria that survive established food processing methods and appear in new food sources. *Listeria monocytogenes*, for example, survives both refrigeration and freezing; and *E. coli* and *Salmonella enteritidis*, commonly considered contaminants of meat and poultry, have recently been found in foods as varied as ice cream, apple cider, and acidic fruit juices such as orange juice, melons, and lettuce.

CHAPTER 13

STINGS AND BITES

||||||||||||||||||||||||||||||||||||||||||||||||||||||||||||||||||||||||||||||||||||||||||||||||||||||||||||||||||||||||||||||||||||||||

Stings and bites can generally be divided in two categories—venomous and non-venomous. Venomous stings usually come from bees, wasps and hornets. Venomous bites come from snakes, fire ants and several spiders including black widow, brown recluse and tarantulas (La Spina, 2009).

Signs and symptoms

- Generalised urticaria
- Itching
- Malaise
- Anxiety
- Laryngeal oedema
- Severe broncho-spasm leading to difficult respiration
- allergic reaction
- presence of hives on site,
- swelling of lips, tongue and/or face,
- Shock
- Death

In general, when the time between the sting and the onset of severe symptoms is short, the prognosis is likely to be worse.

Treating Bee, Wasp and Hornet Stings

Some people have extreme sensitivity to the venoms of the *Hymenoptera* (the stings of bees, hornets, yellow jackets and wasps). Venom allergy is said to be IgE-mediated reaction and thus constitutes an acute emergency. Stings of the head and neck are especially serious, although stings in any area of the body can result in anaphylaxis.

- Using a scraping motion with a fingernail, dull knife or credit, scrape back and forth to remove the stinger. Do not pinch or squeeze, as this will release more venom into the blood stream of the victim and cause a spread of the venom.
- Wash the area several times with soap and water to minimise venomous substance at the site of the sting.
- Apply ice or a cold compress to reduce swelling and relieve the pain. Slight swelling for the following day or two is not uncommon.
- Stings around the face, mouth and throat may require follow-up with a physician.

Emergency management

- Reassure casualty
- Place in a suitable position
- Administer aqueous epinephrine as indicated. Massage the site to hasten absorption. If the sting is on the extremity, apply a tourniquet with sufficient compression to occlude venous and lymphatic flow.
- Remove stinger with one quick scrape of the fingernail. Do not squeeze the venom sac; this may cause additional venom to be injected.
- Cleanse the area with soapy water and apply ice (if available)
- Apply a tourniquet proximal to the sting
- Administer an antihistamine if available
- Treat for shock (anaphylactic shock), if any
- Report to the nearest health care facility for further examination and care

Scorpion sting

Scorpions usually live in hot, dry places. All scorpions produce poison, but this poison vary according to the type of scorpion and its effect on human life.

Signs and symptoms

- Pain at the site, the abdomen, the chest
- Swelling
- Difficulty in breathing
- Eye blurriness

Treatment

- Administer aqueous epinephrine
- Apply ice or ice water on the site (the cold slows down the action of the poison)
- Infiltrate site with xylocaine or chloroquine which act as local anaesthetic agents
- Seek medical attention as possible

Bites

Figure 9 A spider injecting venom to its victim

Treating Spider Bites

Most common spider bites do not cause serious problems. This includes tarantulas. A tarantula bite only causes a mild local reaction in humans.

- Wash the area with soap and water several times a day until the skin is healed.
- Apply antibiotic ointment to prevent infection.

The Black Widow spider and the Brown Recluse spider do require emergency first aid in the hospital. If there is a suspicion that a Black Widow or a Brown Recluse caused the bite, even in the absence of symptoms, medical treatment is required immediately.

Black Widow Spider Bite

The bite of this shiny black spider is rarely fatal, although a life-threatening reaction should not be overlooked, especially in the elderly and very young children. Initially, a burning pain at the site may be reported. This may be followed, in the next 2-12 hours, by limb paralysis, muscle spasms, muscle pain, and abdominal pain in those people bitten on the lower extremities. Children are more severely affected, due to their much smaller size. Nausea, fever, chills and vomiting can also occur.

Prompt medical treatment should be sought for these symptoms.

Brown Recluse Spider Bite

Many have trouble recognizing this brownish coloured spider with a violin-shaped marking on its back. Reactions to the bite of the brown recluse range from mild to severe. Always consult a physician in the event of a Brown Recluse spider bite.

There is generally minor pain at the site at first, but over the next several hours and days, redness or bruising occurs. The centre of the site may form a deep blister in which skin grafting may be necessary in the future. There is no antivenom for the brown recluse spider bite.

Non-Venomous Stings and Bites

Non-Venomous stings and bites generally come from mosquitoes, flies and fleas. Usually these bites require no special treatment. An over-the-counter antihistamine may be useful for itching at the site of the sting. Watch for signs of infection. If the site seems to worsen or if other symptoms develop, consult a doctor.

With prompt treatment most venomous and non-venomous stings and bites can be effectively managed at home or in a medical setting.

Figure 10 A non venomous spider on a victim

Snakebites can be dangerous and fatal.

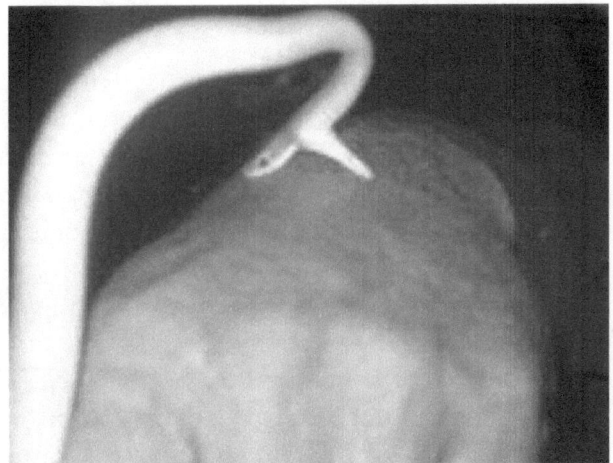

The severity depends on whether the venom has been injected directly into a blood vessel or otherwise and whether or not the snake has a full sac of fresh venom or has bitten before.

Signs and symptoms

- Painfulness
- Swelling
- Nausea
- Vomiting
- collapse

Treatment

- Reassure victim and relatives around
- Absolute rest is very essential (this slows the distribution of the poison)
- Wash wound copiously with water
- Bandage the limb to restrict the venous blood flow from the area
- Remove bandage every half hour for a few minutes and completely discard after three hours

- Administer analgesics such as aspirin, aminopyrin or analgin to relieve pain
- Provide copious beverages (milk, tea, water, soft drinks). Avoid alcohol completely!
- Give artificial respiration and cardiac massage when breathing is laboured and when there are indications of cardiac arrest.
- Seek medical attention immediately an emergency tracheostomy in the case of a swelling of the larynx obstructing respiration and for the administration of antivenin polyvalent serum injection. The administration of the antivenin is either done in line with the *Bezredka method* or giving IV in 500-1000 ml normal saline solution (it should flow for about 4-6 hours) in order to prevent any anaphylactic shock. The serum is administered in fractional doses: first 0.5ml is injected. If no reaction half of the remaining dose is introduced 30 minutes later, and the remainder is injected after another 30 minutes. Antivenin is effective if administered within 12 hours of the snake bite

BIBLIOGRAPHY

Blackman, J.A. 2009. "First Aid." Microsoft ® Student 2009 [DVD]. Redmond, WA: Microsoft Corporation, 2008.

Buyanov, V.M. 1985. *First aid*. Moscow. Mir Publishers.

Busch, P.S.1976. *Living things that poison, itch, and sting*. USA. Walker Publishing Company, Inc.

Chelgren, M. 1980. *Caring for persons with wounds*, edited by Sorenson and Luckmann. London. W.B. Saunders Company.

Cleaver, B., Crawford, R., and Armstrong, V. J. 2006. *First aid manual*. London. Darling Kindersley Limited.

Groot. G.S. 1981. *The role of bandaging in the healing process*. Vianen: Spruyt-Hillen.

Groot. G.S. 1983. *The Relationship between injured and healed*. Vianen: Spruyt-Hillen. 1983.

Kasner, K. and Tindall, D.H. 1984. *Baillière's Nurses' dictionary (20th ed)*. London. ELBS/ Baillière Tindall

LaSpina, J. 2009. First Aid for Venomous Stings and Bites. London. St. Ambulance.

Malcolm R. C. 1990. *Moroney's Surgery for nurses*. London. Churchill Livingstone.

Mayo Foundation. 2011. Harvard guide for medical emergencies. New York. MEMER

Norton. B.A. and Miller A.M. 1986. *Skills for Professional Nursing Practice*. Norwalk. Appleton Century Crofts.

Robroek, W.C.L. and Van de Beek, G. 2009. *Bandages and Bandaging Techniques*. Maastricht. Mediview BV.

Smelter, S.C. and Bare, B.G. 1992. *Brunner and Suddarth's Textbook of medical and surgical nursing* (7th ed). New York. J.B. Lippincott Company.

Stanley, M. 1970. *Baillière's Handbook of First Aid*. USA. Baillière Tindall & Cassell Ltd.

Westaby, S. 1985. *Wound care*. London. Johnson-Johnson Ltd.